The A3 Workbook

Unlock Your Problem-Solving Mind

Daniel D. Matthews

CRC Press
Taylor & Francis Group
Boca Raton   London   New York

CRC Press is an imprint of the
Taylor & Francis Group, an **informa** business

A PRODUCTIVITY PRESS BOOK

Productivity Press
Taylor & Francis Group
270 Madison Avenue
New York, NY 10016

© 2011 by Taylor and Francis Group, LLC
Productivity Press is an imprint of Taylor & Francis Group, an Informa business

No claim to original U.S. Government works

Printed in the United States of America on acid-free paper
10 9 8 7 6 5 4 3 2 1

International Standard Book Number: 978-1-4398-3489-3 (Paperback)

Library of Congress Cataloging-in-Publication Data

Matthews, Daniel D.
   The A3 workbook : unlock your problem-solving mind / Daniel D. Matthews.
      p. cm.
   Includes index.
   ISBN 978-1-4398-3489-3 (pbk. : alk. paper)
   1. Problem solving. 2. Problem solving--Problems, exercises, etc. I. Title. II. Title: A three workbook.

HD30.29.M38 2011
153.4'3--dc22   2010032536

Visit the Taylor & Francis Web site at
http://www.taylorandfrancis.com

and the Productivity Press Web site at
http://www.productivitypress.com

Contents

Preface

My journey as an instructor began while serving in the United States Air Force in the early 1980s. After leaving the U.S. Air Force, I worked for 5 years as a contract instructor and developer at Toyota Motor Manufacturing Kentucky (TMMK). I worked with a small group of contract instructors and developers under David Verble's tutelage. We were charged with developing the course materials that would be used to teach the Toyota Problem-Solving process and the A3 approach to team members at every level in the organization.

After leaving TMMK, I worked for 9 years at Toyota Industrial Equipment Manufacturing (TIEM). While working at TIEM, I had many responsibilities but continued to develop my A3 Problem-Solving skills and the skills of other people.

I eventually went to work for the Manufacturing Extension Partnership (MEP) in Kentucky. I soon realized the power of the A3 Problem-Solving process as I worked with various manufacturers to develop Lean Manufacturing Practices within their organizations. I found that, by following the thought process standardized on the A3 Problem-Solving format, I could quickly and effectively help Kentucky manufacturers make numerous improvements in their processes.

Due to my experience with the Training Within Industry (TWI) curriculum, I was recruited by the Tennessee MEP to conduct Train-The-Trainer classes with their staff. While teaching Job Instruction at one of their clients, I met Todd Shadburn. Knowing of my experience with Toyota, Todd sent a brief e-mail asking me if I had any materials that could explain how to complete an A3.

My reply was simple: "You're in luck." I explained my role with A3 Problem Solving at Toyota in Georgetown. Todd asked me if I could meet with him and give him a quick overview. At the end of our meeting, his only comment was, "None of the books that I've read on A3 have explained the process in a way that is so easy to understand." He has since invited me back several times to teach A3 to team leaders, group leaders, and managers at his facility.

After my discussion with Todd, I realized that although there are some very good books on the market that discuss the A3, there was nothing that walked people block by block through the A3 format and how the problem-solving process fits in with each block. I also wanted to create a workbook that could be used at every level in the organization to develop the basic problem-solving skills that are required to make continuous improvements.

My hope is that this workbook will be used by companies to develop the problem-solving skills of their employees—leading not only to improved profit, quality, productivity, safety, and delivery, but also to a culture that understands the value of developing people at every level.

TWI job relations stresses utilizing people to the best of their abilities. By teaching A3 Problem Solving to those who actually accomplish the work, whether on the shop floor or in the office, organizations will be able to solve or reduce the severity of most problems.

The text in this workbook is structured to follow the layout of a basic Problem Solving A3 format. I wrote it in this format so that the reader could actually practice applying the skills described in each section. There are case studies that the readers will use to complete A3s, tips on how to improve the readability of A3s, examples of Problem Solving A3s, and a proposal A3.

An effective problem-solving process is a critical part of implementing efficient business practices. A problem-solving culture is a fundamental component of empowering employees to support business improvements. Both of these components, taken together, can help each and every organization make continuous improvements on the long journey to creating a more productive and profitable business.

Acknowledgments

There are too many people for me to recount who made this journey possible. It would be like listening to one of those award speeches that goes on and on. However, I would like to thank a few key people who directly contributed to making this workbook a reality.

I need to thank David Verble for including me on the team that developed and taught A3 Problem Solving at Toyota Motor Manufacturing Kentucky (TMMK). Later he took me under his wing as he branched out into Practical Problem Solving. Because of David's mentoring I was able to leave TMMK and take a permanent position at Toyota Industrial Equipment Manufacturing (TIEM). I spent almost 10 years at TIEM applying David's teachings and insight. I can never thank David enough for what he did for me and for my career.

For providing me with insight into the Japanese way of conducting business and the use of Namiwashi, I would like to thank Mr. Imaeda and Mr. Mizuno, my Japanese coordinators at TMMK and TIEM.

I would like to thank Todd Shadburn and Keith Groves, Jack Parsons, Brent Renfroe, and Lynn Witten Godsey for their support.

A special thanks goes to my wife Mart, good friend James R. Johnson, and my dad David for reading my manuscript and providing me with valuable feedback. Finally, I would like to thank my family for their support over the years.

About the Author

Included in **Daniel Matthews'** 30-year career is more than 14 years of supervisory and management experience with Toyota Motor Manufacturing and Toyota Industrial Equipment Manufacturing. He is skilled as a trainer, coach, and implementer of Lean Manufacturing, having trained hundreds of associates in the methods of the Toyota Production System (TPS).

During his time with Toyota, Daniel became an experienced Training Within Industry (TWI) instructor. The TWI program is widely recognized as the foundation for Lean Manufacturing.

Daniel has helped both leaders and associates build the skills they need to support a Lean culture, including problem solving, team building, facilitation, coaching, communication, conflict management, and leadership both in the classroom and on the shop floor.

Daniel's skills as a TWI and A3 Problem Solving instructor have led to speaking engagements in the printing, automotive, and general manufacturing industries.

Daniel graduated summa cum laude from Indiana Wesleyan University with a degree in business administration where he was a two-time recipient of the Outstanding Business Professional Award.

Chapter 1

Introduction

The Origin of A3

What is an A3 format? First and foremost, A3 is a format developed by Toyota for telling the story of improvement. The A3 has two basic functions, one as a method for making proposals and the other as a means of reporting on the approved actions as outlined in the A3 proposal.

In actuality, the A3 format earns its name from the International Organization for Standardization's (ISO) designation for paper measuring 297 by 420 millimeters. This is the paper size that has become the standard by which Toyota communicates continuous improvement projects. The A3 paper size used at Toyota Motor Corporation in Japan measures 297 by 420 millimeters (or 11.793 by 16.535 inches). This paper standard is used in just about every country in the world other than the United States. In the United States, paper size is based on the American National Standards Institute (ANSI) paper standard (Table 1.1 compares ANSI and ISO paper sizes).

ANSI B (11 by 17 inches) is the closest match to the ISO A3 paper size and therefore is commonly referred to as A3 by Lean practitioners and those familiar with Toyota's A3 format. Because ANSI B Problem Solving does not roll off the tongue as eloquently as A3 Problem Solving, I will continue to use the term "A3" throughout this book. At Toyota in North America, the Japanese coordinators continue to use ISO A3 paper. ISO A3 paper provides an additional 6 square inches of space for telling the story of improvement. The goal is for you to be able to communicate your proposal or problem on a single sheet of paper. ANSI B paper and the A3 Problem Solving process will make it possible for you to achieve this goal.

The idea behind the A3 is simple: communicate your proposed idea on a single sheet of paper—no more and no less. In a Lean organization where everyone has multiple functions, there is little time for reading reams of data to understand a particular problem or situation. The A3 effectively condenses large amounts of

Table 1.1   ANSI and ISO Paper Sizes

INCHES			
ANSI Sizes		ISO Sizes	
ANSI A	8.5 × 11	8.3 × 11.7	A4
ANSI B	11 × 17	11.7 × 16.5	A3
ANSI C	17 × 22	16.5 × 23.4	A2
ANSI D	22 × 34	23.4 × 33.1	A1
ANSI E	34 × 44	33.1 × 46.8	A0

data into an easy-to-read and understand format. By keeping it simple, you are less likely to lose the attention of the reader and possibly their support.

Although you do not need to create an A3 for every situation, it is a good idea to use the format on a regular basis. The more you use it, the more it will become a natural part of how you approach problem situations. In your daily work, you will encounter many situations requiring action. Not all situations will require the creation of an A3, but the thought process can be used at any time.

There are several documented benefits to using the A3 Problem-Solving approach, as summarized below:

- Provides a methodical approach to problem solving
- Provides a succinct format for presenting or reporting facts to others
- Documents a trail that others can follow and use to understand the problem solver's actions and results
- Provides a common language and method within an organization
- Creates a culture conducive to sustaining Lean Manufacturing concepts
- Provides a foundation and lays the groundwork for future change

A3 Formats

As discussed earlier, A3 is a format created by Toyota for telling the story of improvement. The A3 has two basic functions: first as a method for making proposals, and second as a means of reporting on the approved actions as outlined in the proposal.

By condensing the essential information to fit on one page, the A3 format makes it easier for anyone in the company to read and understand what the author is proposing or reporting. The A3 methodology is also used to mentor subordinates on how to become problem solvers, not problem bringers.

The A3 Proposal Format is used when a Team Member requires management approval to make a change or to head off any anticipated problem. The blocks of

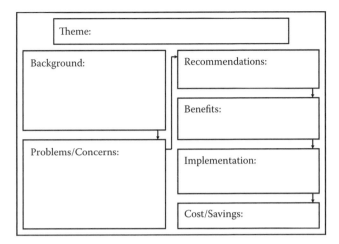

Figure 1.1   Possible A3 proposal format.

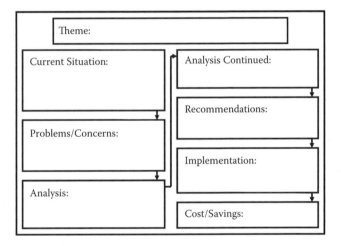

Figure 1.2   Possible A3 proposal format.

the proposal format can have many different headings; Figures 1.1 and 1.2 represent two possible configurations for an A3 Proposal. Appendices T and U are examples of completed A3 Proposal Formats that can be used for future reference.

The A3 Problem Report is used when a Team Member requires management approval for implementation of countermeasures to eliminate an existing problem. The A3 Problem Proposal/Report Format is both a proposal and a report. Initially it is a proposal that is presented to management and must be approved before implementation can begin. It becomes a report when the owner begins to see results from the countermeasures and reports those results to management. Once the Problem Report is mastered, creating other A3 forms becomes much easier; Figure 1.3 is the Problem Report Format. Appendix S is an example of a completed A3 Problem Report Format and Appendix V is a blank A3 Problem Format that can be reproduced.

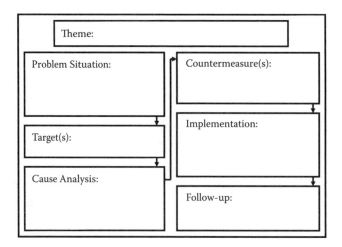

Figure 1.3   A3 problem report format.

Folding the A3

Because the standard A3 and ANSI B paper is larger than the paper used for day-to-day writing and reporting, it was found difficult to file and incorporate into report binders. For this reason, Toyota adopted a specific way (standard) for folding the A3. By folding the A3 in half from right to left, you now have a sheet of paper in its folded state that is 8.5 by 11 inches. The opening will be on the left side if folded properly. Then, by taking the top edge on the left side and folding it evenly back to the right, you have a crease on the right side.

By folding the A3 in this accordion-like manner, you are able to file it more easily. In addition, you are able to place it in report binders that contain supporting information for the A3. With the top edge facing to the right, you have a natural tab that can be grasped and pulled, thus making it easier to open and view the contents of the A3. Figure 1.4 depicts how to fold an A3.

How A3 Fits into Your Organization

The purpose of this A3 workbook is to provide anyone at any level within an organization with the tools needed to be an effective problem solver. More and more, managers are realizing that they must develop their entire workforce in order to help their organization achieve its goals and objectives.

In a meeting with a manager from an automaker, the topic of problem-solving tools, in particular Six Sigma, came up. He told me about a study conducted by one of his Original Equipment Manufacturers (OEMs) regarding the types of problems they experienced. The problems were categorized in the following three ways:

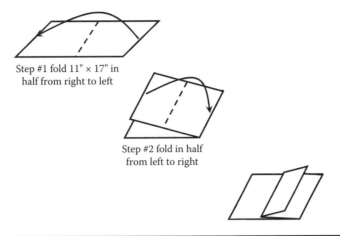

Figure 1.4   How to fold an A3.

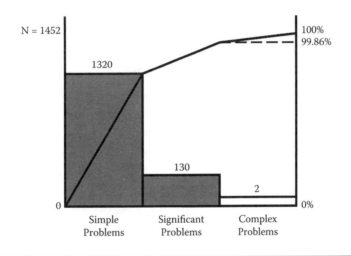

Figure 1.5   Pareto of problem categories.

1. Simple (90.91%): everyday problems that require basic problem-solving skills
2. Significant (8.95%): major problems that require supervisor or management buy-in and approval before implementation
3. Complex (0.14%): problems that may require more complicated methods such as Six Sigma

Based on the study, 99.86% of problems experienced at the OEM were easily handled at the tactical or Team Member/Team Leader level; Figure 1.5 is a Pareto chart showing the three problem categories. The problem is that most organizations do not develop these basic problem-solving skills throughout their company. At Toyota, we taught problem solving in conjunction with the A3 to all departments and all levels. The A3 Problem-Solving process is one that is time efficient, can be driven down to the process level with little training, and has a high effectiveness after writing just a few A3s.

With Six Sigma, a great deal of training is needed in order to gain the knowledge necessary to work on a project. The nature of most Six Sigma projects requires months of training and months to complete a project.

Most Six Sigma projects should save in the neighborhood of $150,000 to $500,000. According to the OEM study, if a company experienced 1,000 problems during the course of a year, only one would be categorized as complex enough to require Six Sigma tools. That means that a company could realize a savings of $500,000 from that one project.

If an organization trained all its workers in basic A3 Problem Solving and tackled the remaining 999 problems resulting in an average savings of $1,000 per problem, then the company could realize a total savings of $999,000. I have worked with companies that have realized millions in savings as a result of A3 Problem Solving. In one situation during a Kaizen event, I used the process to help the group identify the problem and get to the root cause quickly. It took less than 45 minutes and resulted in $2.25 million in increased revenue without overtime.

It is not about making a decision to use Six Sigma or A3 Problem Solving; rather, it is about using the best tool for the job. To be an effective golfer, for example, you need a driver for those long shots and a series of irons, including the putter, to get you progressively closer to the hole. A3 and Six Sigma are both useful tools that can help a company generate a great deal of savings. However, it is important to use the right method at the right time. Using Six Sigma on simple or significant problems when A3 Problem Solving is a better fit is like using a driver to sink a 2-foot putt!

The A3 process can be easily taught and practiced by every level in the organization. This creates an organization of problem solvers that can address more than 99% of all problems encountered in any organization. In some cases, A3 projects may produce smaller incremental rewards, but organizations will gain greater cumulative improvements and savings through A3 Problem Solving, not to mention improving the cognitive skills of all employees.

Chapter 2

Overview

A3 Problem-Solving Overview

The A3 Problem-Solving Report Format is broken down into seven distinct blocks. Each block fulfills a specific function in the process of solving problems. The first five blocks help the problem solver create a PLAN for successful problem solving.

- Block #1—Problem Situation:
  - Background
  - Standard
  - Current Situation
  - Discrepancy
  - Extent
  - Rationale
- Block #2—Target:
  - What to do
  - When to do it
  - How much
- Block #3—Theme:
  - General theme categories
  - Creating the theme
- Block #4—Cause Analysis:
  - Potential Causes
  - How Check
  - Results
  - 5-Why Causal Chain
  - Root Cause

■ Block #5—Countermeasures:
   – Short term
   – Long term
   – Why Recommended

The sixth block of the A3 provides a schedule to DO what needs to be done to get the desired results and a section to CHECK the progress of the plan and its effectiveness.

■ Block #6—Implementation:
   – What actions need to be taken
   – Who should take each action
   – When each action needs to be completed
   – Results of actions

The seventh block of the A3 is designed to identify how the problem solver is going to CHECK to see how effective the countermeasures are in relation to the discrepancy. It also explains what ACTION the problem solver will take once results are documented.

■ Block #7—Follow-up:
   – How Check
   – When Check
   – Recommendations

A3 and the PDCA Cycle

The Plan-Do-Check-Action (PDCA) cycle that most people are familiar with depicts a circular thought process with four equal quadrants. Figure 2.1 represents the typical PDCA visual.

After teaching A3 Problem Solving for 20-plus years and having helped other organizations with their problems, I have refined my visual representation of the PDCA process. Although all four quadrants of the PDCA process are important to the success of any problem-solving effort, the Plan phase by far requires more time and effort. Creating a good plan does not happen by accident. You must first take the time to identify the problem precisely. Then you must analyze the problem to root cause so that you can select the best countermeasures for the situation.

Like problem solving, cooking a great pot of chili from scratch has four distinct phases. First you have to spend a considerable amount of time *preparing* (PLAN) the contents: chopping meat and vegetables, roasting chilies, soaking the beans, and measuring out the spices. The next phase is to *combine* (DO) all the ingredients in the pot at the right time to simmer so that all the textures are just right when you are ready to serve the chili. After everything is combined

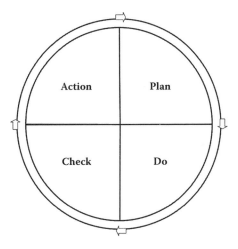

Figure 2.1   Typical PDCA visual.

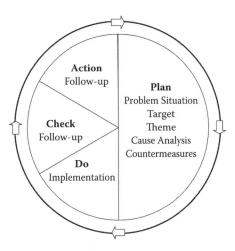

Figure 2.2   Modified PDCA cycle.

and simmered for the appropriate amount of time, you have to *taste* (CHECK) the chili to see if it meets with your expectation. Finally, you need to *decide* (ACTION) if the chili meets your expectations. If it tastes right, you standardize the recipe so that you can replicate the chili. If it is too mild, you may want to add some more spices before serving. If it just does not measure up, you may scrap the chili and start all over again.

Preparation of a great pot of chili requires a great deal of hands-on work, just like the Plan phase of the problem-solving process. Combining all the ingredients, tasting the chili, and deciding if the chili meets your expectations require far less hands-on work and more monitoring and decision making, just like the Do, Check, and Action phases of the problem-solving process. For that reason, I find that the modified PDCA cycle (Figure 2.2) is a more accurate depiction of the time required for each phase in the A3 Problem-Solving process.

We've all heard the expression "you must go slow to go fast." That is exactly what I try to communicate to my clients when they become frustrated by the process. I often find myself having to draw a modified PDCA cycle to accurately depict the time required to create an effective plan. This modified PDCA visual helps me maintain their focus and allows them to see how they are progressing through the A3 Problem-Solving process.

Definition of a Problem

Before starting to explain the A3 process, it is important that we all understand how a problem is defined. We will use the Toyota standard for this discussion. Having this common understanding within Toyota enables all Team Members to communicate situations in an easy-to-understand way.

- ■ Problem definition: A problem is the *difference* between the Current Situation and the Standard (or Standard Way).
  - The Standard is a specific *known* expectation or norm describing what should be happening in a particular situation. If the standard is not known, then it is just one person's perception of the way things should be in a given situation.
  - The Current Situation is a description of what is happening at that point in time as it relates to the standard.
  - The Difference is the measurable or recognizable variance between the standard and the current situation (also referred to as the discrepancy, gap, or problem).

When I ask people what the problem is, they typically start talking about possible causes of the problem, versus explaining how the current situation differs from what is expected. Having this common language keeps people from jumping to conclusions during the initial stages of the problem-solving process.

Types of Problems

Problems can be characterized in one of three ways:

1. *Preventive (Soshi).* This type of problem solving focuses on preventing issues from becoming problems. Issues are likely to occur due to policy changes or other organizational changes, so it is important to be proactive to keep these issues from blowing up into bigger issues that become real problems. In preventive problem solving, you anticipate what problems could arise due to changes and take actions to prevent the problems from surfacing.
   A proactive problem-solving example follows. A company currently gives every employee a Christmas bonus check at the end of October. The company has decided to eliminate the bonus and go to a quarterly incentive

program. The average quarterly incentive will have an overall higher pay-out than the Christmas bonus being discontinued.

The company feels that most, but not all, Team Members will think that getting four incentive checks per year will be better than one bonus check. To avoid problems, management would need to look at all the potential reasons workers may find fault with the new program and defuse any issues before the program is announced.

This was an actual problem at one place I worked and there were a few Team Members who opposed the new plan. If management had anticipated the problem, they could have avoided the outbursts that occurred during the meeting when the new program was announced for the first time. This proactive approach is a key learning in the TWI Job Relations class. One of the four foundations of good job relations is to tell people in advance about situations that will affect them.

2. *Continuous Improvement (Kaizen).* Problem solving focused on improving an existing program, system, or process (Figure 2.3) represents the process of Kaizen. After making an improvement, it is essential to have the ability to sustain that effort for a given period of time to demonstrate stability in the process. Once process stability has been established, Kaizen can begin again.

The process begins with defining the new standard or expectation, then progresses through analyzing the obstacles, developing and implementing measures to address the obstacles, and following up to ensure successful achievement of the new standard. Then the cycle repeats once stability of the process has been established.

This is a huge part of Toyota's culture. Team Members at every level are encouraged to see how they can make incremental improvements to their processes through quality circles and the suggestion system. These incremental improvements may seem trivial to some organizations, but it is the cumulative effect that these incremental improvements have over the years that has enabled Toyota to succeed.

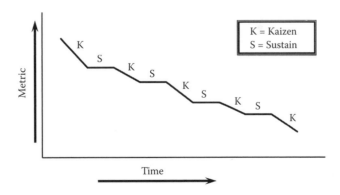

Figure 2.3   Kaizen process.

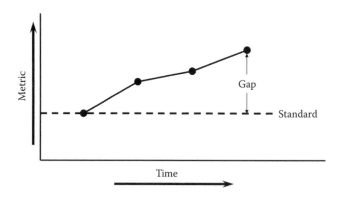

Figure 2.4　Iji type problem.

3. *Maintenance (Iji)*. The third type of problem solving addresses situations where an established standard is not being met. Everyone has found himself or herself in a situation where things suddenly or gradually get off track from the norm. In these situations it is essential to get back on track with minimal disruption. The sooner the problem is solved, the less impact it will have on the organization. The majority of this text focuses on this type of A3 as it is the most difficult to master. Figure 2.4 represents an Iji type problem.

How Problems Come to Us

Problems are typically identified in one of four ways:

1. *The problem is sensed.* If you sense that a problem exists, it is usually because your sense of what the normal situation should be and the way you are seeing the current situation do not match up in your mind. At the Team Member level, the initial sense of the problem may be something as benign as having to periodically pick up boxes that keep falling off the roller conveyor. If a Team Member senses the problem at this level, other more serious problems can be avoided. If the Team Member does not recognize and fix the problem at this level, it could lead to delays in production due to damaged product.
2. *The problem is picked up.* Picking up a problem occurs when someone compares available data to a specific standard, norm, or expectation, usually based on metrics that support achieving critical organizational goals. These metrics are generally referred to as KPIs (Key Performance Indicators) in Lean Manufacturing. KPIs are extremely important to the continuous improvement efforts of Lean organizations. If the Team Member in the earlier example does not sense the problem, the Team Leader may see a decrease in productivity after examining the production results. If the problem does not create a delay in production, the Team Leader may not pick up the problem until after the product is inspected and weekly or monthly quality results are posted.

3. *The problem is passed on to you.* Your Team Leader, Team Members, and other departments within the organization can also pass on problems to you. The Team Leader may not find out about the defective product until after the daily production meeting where Team Leaders from all departments meet to discuss and share issues.

4. *The problem bursts in your face.* The final and worst way a problem can come to us is when it bursts in our face. This could be in the form of an accident, customer complaint, recall, or some other serious situation that arises. If the seemingly non-issue of boxes falling off the conveyor is not sensed, picked up, or passed on, you may experience a problem that bursts in your face in the form of a large order that is returned due to damaged parts.

Chapter 3

Problem Situation

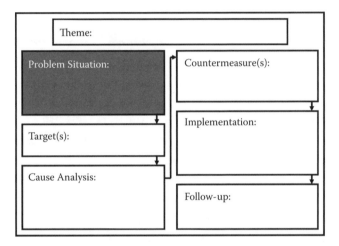

Figure 3.1 A3 problem report. The problem situation is the second block of the A3, but it is the beginning of the process for creating an A3.

Introduction

The problem situation (see Figure 3.1) is the second block of the A3 but it is the beginning of the process for creating an A3. The problem situation consists of:

- Background
- Problem Statement
  - Standard (expectation or norm)
  - Current Situation (what is happening now)
  - Discrepancy (gap or problem)
- Extent
- Rationale

Background

When starting the problem-solving process, think about what your level or position is in relation to the organization. By doing this, you are able to provide the reader with background into your situation and how this problem is relevant to the organization as a whole:

- What area do you work in?
- What function is your area responsible for?
- What is your position within the area?
- What work do you perform?
- What, if any, relevant history surrounds this situation?

Briefly explain to the reader what area or department you work in and what specific function the area provides for the organization. You will also need to provide a description of your position and the responsibilities you have in that position.

As a problem solver, it is also a good idea to provide information about events or situations that may be affected by or have an effect on the problem you have picked up. You may not become aware of situations that affect the problem you have selected until you are deeper into the process. This is one of the reasons the Japanese encourage you to write A3s in pencil, because you may have to make numerous changes to the A3 as you learn more about the situation.

Problem: 20% of dashboard covers coming out of molds 3 and 5 on the dashboard carousel are missing the back right corner.

Background example: I am the Team Leader of the dashboard cover carousel area within the plastics department of Lotta-Lift. I am responsible for identifying and eliminating non-value-added work from the process. Carousel production is scheduled to increase beginning in August due to the increased demand for 1.5-ton lifts.

Standard

It was not until I left Toyota and began working with other manufacturers that I truly understood how intertwined all the components of the Toyota Production System (TPS) are, and especially how the Problem-Solving component plays a major part in every aspect of TPS.

If you look at every component of TPS, you begin to see how standards play a part in everything Toyota does. Standards are everywhere. For instance, once a Team Leader creates a Job Instruction Job Breakdown, it becomes a standard that can be compared to the way a Team Member performs a given job or process. Each major step and key point in that process is defined and can be audited to see if there is any discrepancy or deviation. When a problem is identified and tracked back to a specific process, the Team Leader can use the Job Breakdown Sheet to evaluate the Team Member's ability to perform the task properly.

Likewise, Toyota's 5S system creates standards that make it easy for Team Members to tell if the process has all the required items to do the job, like materials, tools, and fixtures. Good 5S practices make the abnormal condition obvious to anyone, making it easier to see when there is a problem.

Another use of such standardization is the Kanban. The Kanban specifies how many parts are to be delivered to a specific location. A Kanban procedure typically dictates that the Kanban be placed in the Kanban mailbox when a parts box is first opened. If the Team Member does not place the Kanban in the mailbox or does not pull the Kanban until the box is empty, parts will not arrive on time.

Even the well-known Andon (a cord that a Team Member pulls to stop the line when a defect is detected) is based on a standard. The Team Member has a certain Takt time or cycle time to perform a given task. If the Team Member makes it to the 75% mark and has not completed 75% of the standard work, the standard may dictate that the Team Member pull the Andon. The Team Leader responds to the Andon and tries to fix the problem before the product reaches the 100% point, at which time the line will stop. This Andon standard prevents problems from being passed on to the next process (Figure 3.2 represents a process area based on standards).

Standards can also be classified as expectations and norms. In situations where there is no defined standard, there is most certainly an expectation or norm that will apply. Without this baseline, it would be impossible to understand the magnitude of the perceived problem, much less begin the process of solving the problem. This culture based on standards makes it easy for all levels within the organization to know when there is a problem. Another component is to know how to solve

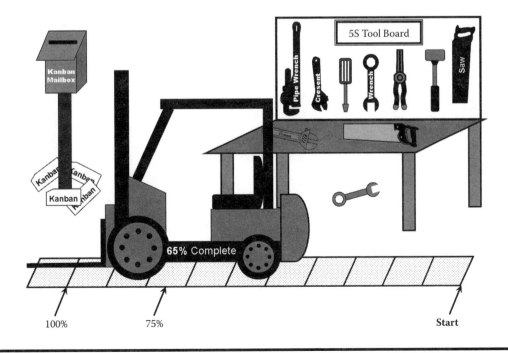

Figure 3.2 Production standards.

problems and to know that you have the power and responsibility to do so. This is the purpose and power of the A3 Problem-Solving process.

By developing a manufacturing process based on standardization and developing problem solvers at every level, Toyota has been able to outpace the productivity and efficiency improvements made by most American firms on a year-over-year basis. The foresight to develop the problem-solving skills of all its Team Members has led to Toyota's growth in the United States and abroad.

I was not aware of how completely the A3 thought process affected my daily activities until a co-worker pointed it out to me. During various events where we were paired together, he noticed that I always asked the same question every time someone asked me what to do in a given situation. My first response is to always ask, "What is the standard or expectation?" My co-worker thought I was just buying time until I could come up with a good answer to the question. However, if someone is asking me what action should be taken, I have to know what the expectation is before I can provide an informed response. Usually after asking this question and getting a response from the person, it becomes obvious to the person what should be done.

Knowing the standard is critical to the process of deciding whether or not a problem exists. Standards benefit Team Members by providing a baseline to measure their progress against. This allows Team Members to be proactive and notify their Team Leader or Supervisor at any time during the day if an expectation is not being met.

In the A3 Problem-Solving process, there are standards for a standard. A standard should:

- Be specific as to the expectation or norm
- Be tangible and recognizable
- Be quantifiable and measurable
- Be shared and agreed to

The following are examples of standards that meet the criteria of a clear standard:

- Each Team Member must trim and package six turkey thighs every 60 seconds for a total of 60 (six packs) per hour.
- Each Team Member must produce a total of 5,000 boxes of #6 nails each shift on each machine 1, 2, and 3.
- Each operator on line #7 should assemble and test one circuit board every minute.
- The receptionist should have all mail metered and ready for security to pick up by 3:30 p.m. Monday through Friday.

- The Human Resource specialist must have payroll completed, checked, and submitted to the corporate office by 10:00 a.m. Eastern Standard Time (EST) every Monday, or the Tuesday after a Monday holiday.
- Nurses must wash their hands upon entering and before leaving each patient's room.

In a successful Lean organization, the company must establish specific standards or objectives and drive them down to every level in the organization, including the Team Member level. It is imperative that organizations looking to make a Lean transformation develop a culture based on standards.

Many of the companies that I visit have overall metrics for how they want their company to perform. However, few have taken the time to drive these metrics to the Team Member level. Fewer yet have developed the necessary problem-solving skills required in their people to efficiently achieve the company's overall goals or objectives. The main reason organizations fail to sustain Lean and other initiatives is due to the lack of shared standards and the expectation that these standards be achieved.

It is extremely important for management teams to communicate what their priorities are and how these priorities will steer the organization to success. If there is no standard, clarifying the problem becomes more difficult. Without a clearly defined problem, it will be difficult to create effective countermeasures that lead to improvement. To clarify the problem, you will first need to determine the standard for the situation.

In a situation where boxes keep falling off of a roller conveyor, what is the standard? Is it acceptable for boxes to fall on the floor? Is it OK that the operator spends time picking up boxes? Because picking up boxes is a non-value-added activity, a Team Member with a problem-solving mind would begin to clarify the expectation or standard. The Team Member would need to look at the standardized work for the process to find answers to these questions.

Standardized work will provide the work sequence (standard based on the way to do the job) and the Takt time (standard based on a level or amount of time to do the job) for the process. If the operator is exceeding the Takt time, then it is a problem. If picking up boxes that fall off the conveyor is not part of the work sequence, then this is also a problem.

There are two types of standards:

1. Standard levels: provide a specific time or quantity that must be achieved.
2. Standard ways: describe how something must be done to achieve the desired result.

There are standards that are quantifiable, like having a defect rate of no more than 0.0005 on cabinet hardware requiring secondary buffing. There are also standards that can be counted and are based on how something is done, such as specifying that suppliers practice first-in-first-out (FIFO) inventory control practices when making hourly deliveries to the facility. Both types of standards are measurable and specific.

Production Standards Exercise

Read the following examples and circle the example numbers of those with clear standards and convert the ineffective standards to effective standards. Also categorize each by type (standard level or standard way). See Appendix A for answers to this exercise.

Example 3.1

It should only take a couple of minutes to tap the holes for the door hinges.

Example 3.2

Team Members working the fill station must fill 100 bags of feed every hour.

Example 3.3

All Team Members on the 5S Kaizen team should participate in meetings.

Example 3.4

Team Members must place Kanbans in the Kanban mailbox as soon as the first part is pulled from each box.

Example 3.5

Cartridges should slide into brass collars without having to be tapped into position.

Example 3.6

All welds on the forklift frame must be continuous with no gaps and be between 11 and 11.5 millimeters wide.

Example 3.7

The height of the label should fall between 30 millimeters and the upper limit of the standard.

Example 3.8

It should only take a little while to change from one product packaging to the next.

Office Standards Exercise

Read the following examples and circle the example numbers of those with clear standards and convert the ineffective standards to effective standards. Also categorize each by type (standard level or standard way). See Appendix B for answers to this exercise.

Example 3.9

The Mailroom Team Member should fill all copiers with paper before leaving at the end of each day.

Example 3.10

On the second Monday of every month, all expenses from the previous month should be submitted to the Finance department to be paid on the last Friday of the current month.

Example 3.11

Office staff should conduct regular Quality Circle meetings.

Example 3.12

Payroll staff must have all hours for the previous week entered into the Payware program no later than noon every Monday, or by noon on the first day back after a Monday holiday.

Example 3.13

The 6-month calendar must be approved and posted on all bulletin boards by the last Wednesday in June and the first Wednesday in December.

Example 3.14

The receptionist should never leave the front desk unattended for more than a few minutes at a time.

Example 3.15

Software can only be added to company PCs by the Information Systems department.

Example 3.16

Personal radios should not be turned up too loud.

Current Situation

The first step in clearly identifying the problem is to ignore your natural inclination to take some kind of action. When you "sense or perceive" that there is a "problem," the typical response is to figure out what action should be taken to eliminate the "problem." In the A3 Problem-Solving process, you are encouraged to delay acting on the problem until you have stated it clearly and precisely.

The current situation is defined as *the way things are now.* You will most likely encounter a set of circumstances that is out of the norm. When this happens, it is critical that you identify the facts that explain what is actually occurring and try to make your first "sense" of the problem more specific. To do this, you must remove the subjectivity and replace it with facts. Understanding the current situation begins with the level at which you pick up the problem.

Problems are picked up at different levels within the organization. Figure 3.3 represents the problem perception at different levels. The further you are from the process, the more vague the problem becomes and the more you will need to clarify and break down the problem. This is another good reason to educate Team Members at the process level about A3 Problem Solving.

By developing problem-solving skills at the lowest level in the organization, the company can keep many problems from escalating in magnitude. This workbook primarily focuses on problems at the Supervisor and Team Leader level. Seeing the process from this level will allow you to easily solve problems that are picked up at their most vague or specific level.

Most of the time, our initial sense of the problem focuses on an annoyance such as walking long distances to complete the process. Behind that annoyance, which is usually subjective in nature, there is usually a more important issue that should be addressed.

- *Subjective:* Engineers did not know what they were doing when they set up my process.
- *Fact:* My process takes 20 minutes to complete, and 10 minutes of that time is spent walking and waiting.
- *Subjective:* The equipment I use in my process is too far away.

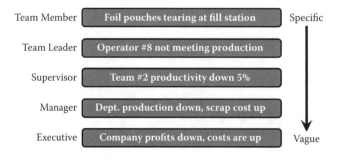

Figure 3.3   Problem perceptions at different levels.

■ *Fact:* The chop saw is 400 feet from my workstation, and the shear is in another building 1,000 feet away. I need both pieces of equipment to build each unit.

To accurately depict the current situation, it is important to look at existing data related to the situation. Make sure that you go and see the situation for yourself; do not rely on others to provide you with the information you need. Problem solving is not done in a climate-controlled office while sitting at a desk. Consider any differences or changes and the timeline of those changes compared to the timing of the problem.

Example

It currently takes the Team Member on process #5 an average of 20 minutes to produce one unit. Actual work time or value-added work takes 10 minutes and non-value-added work takes 10 minutes.

If there are no data, you will have to gather the data. To do this, you may need to create a check sheet and post it in the area so that you can track occurrences related to the perceived problem. Table 3.1 shows non-value-added work for daily production.

The data you gather will also help you later in the process when you have to break down the problem in order to precisely state your problem. As a Team Leader or Supervisor, you may grasp problems using the information posted on your team or department board. Figure 3.4 shows a basic Lean Manufacturing Department Status Board. These boards track information related to safety, quality, training, cost, scrap, delivery, and productivity. The board may also display a future-state Value Stream Map (VSM) of your product along with all the required action items.

The use of department or team boards makes it easy for anyone in the facility to see the status of the area at a glance. At Toyota, management would make several trips around the shop and office each day looking at these boards and asking questions regarding the department status. Having such transparency in the work area makes it difficult to hide problems. This transparency actually drives leadership to continually improve.

Table 3.1   Non-Value-Added Work for Daily Production

Process #5 Non-Value Added	Daily Production Units										Avg.
	1	*2*	*3*	*4*	*5*	*6*	*7*	*8*	*9*	*10*	
Monday	9.0	10.5	12	9.5	10	10	9.5	11.5	9	9	10
Tuesday	9	10.5	10	9	10	13	9.5	11	9	9	10
Wednesday	9.5	10.5	8.5	9.5	10	10.5	9.5	11.5	9.5	11	10
Thursday	10	9.5	12	9.5	12	10	10	9	9	9	10
Friday	9	12	12.5	9	10	8	9.5	11.5	9.5	9	10

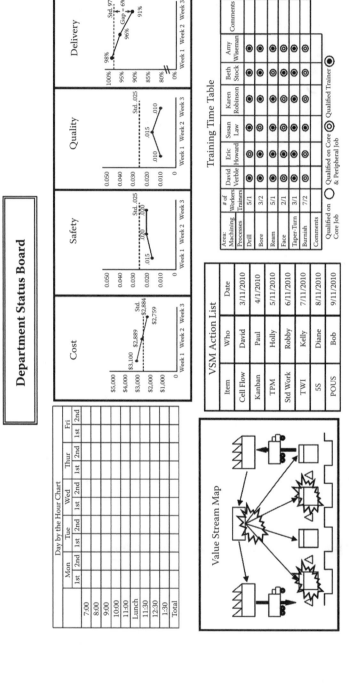

Figure 3.4   Lean Manufacturing Department Status Board.

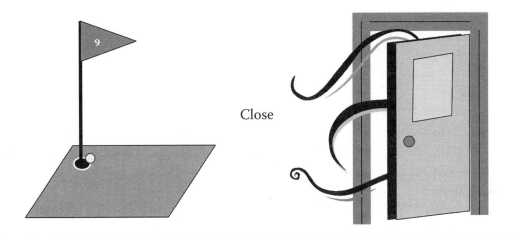

Figure 3.5   How context changes perception.

Regardless of how a problem comes to you, it must be fully grasped in order to begin understanding the true problem. In other words, put the problem in perspective as it relates to where the problem exists within the organization and how it relates to the organization's goals and objectives. Identify the actual problem by taking in all the facts of the situation so that you can see the bigger picture. By looking at the bigger picture, you will have a better understanding of how important the situation is in relation to other issues and how you should proceed.

For example, take the word "close" by itself. The reader has no idea what the writer is trying to communicate. You need to be able to see the word in the same context as the writer. Is it being used as an adjective? Or is it being used as a verb? The context of the word, combined with the text, will let you know how you are to react to the situation. Figure 3.5 shows how context can change your perception. If you are golfing, you may say the ball is close enough to the hole to call it a gimme. If you are a homeowner worried about utility bills, you will want to close the door to keep from wasting energy. As you can see in this example, the context provides you with the necessary information needed to make the appropriate decision. As a problem solver, you need to think about the context of the problem and how it fits with the company objectives.

Discrepancy

A discrepancy is a measurable or recognizable difference between a standard level or standard way and the current situation. The discrepancy should clearly highlight the difference between the standard and the current situation.

Example

- *Standard:* The tank weld Team Member should only take 15 minutes to tack weld the tank together, 1 hour to weld the tank, and 20 minutes to grind all welds.

■ *Current Situation:* The Team Member on the tank welding process takes 30 minutes to tack weld the tank together, 1 hour to weld the tank, and 20 minutes to grind all welds.

■ *Discrepancy:* It is taking the Team Member 15 minutes longer to tack weld the tank.

Determining the discrepancy may seem simple; however, the discrepancy by itself is rarely enough to provide the problem solver with the needed information to begin looking for the root cause. The discrepancy is the starting point for breaking down the problem into specific characteristics that can be analyzed to root cause.

The Standard, Current Situation, and the Discrepancy are combined to create the basic Problem Statement. The following are examples of production and office problem statements. Read each statement and decide which ones represent good examples of the three components of a problem statement. For those that you feel are not good examples, make notes on what would need to be refined. See Appendix C and Appendix D for answers to these exercises.

Production Problem Statement Exercise

Example 3.17

- Frame weld must produce a completed head guard every 25 minutes.
- 92% of the time, frame weld produces a completed head guard every 25 minutes.
- 8% of the time, frame weld takes 27 to 31 minutes to complete a head guard.

Example 3.18

- It should only take a couple of minutes to tap all three door hinge holes.
- It is taking 10 to 16 minutes to tap all three door hinge holes.
- This is five to eight times longer than expected.

Example 3.19

- No more than 2% of faucet handles should need rebuffing.
- 5% of gold faucet handles and 3% of silver faucet handles need to be rebuffed.
- Gold handles exceed the standard by 3% and silver faucet handles exceed the standard by 1%.

Example 3.20

- No more than three injection-molded parts should have color mix after a color changeover.
- Injection mold machines 1, 2, and 4 experience two to three parts with color mix after a color changeover, whereas injection mold machines 5 and 6 experience five to seven parts with color mix after a color changeover.
- Injection mold machine 3 is offline.

Office Problem Statement Exercise

Example 3.21

- All 10 supervisors must have all their Team Member performance appraisals submitted to HR no later than April 15.
- As of April 15, 78 of 96 performance appraisals were turned in to HR.
- Performance appraisals are late.

Example 3.22

- All ink cartridge boxes are to be kept for shipment to the recycler.
- There are seven cartridges that need to be shipped to the recycler and only four boxes.
- Three ink cartridge boxes are missing.

Example 3.23

- Only a limited number of staff members can be off at one time.
- Three purchasing staff members are off this week.
- Other departments are complaining about poor service.

Example 3.24

- The Safety department set a standard requiring all Team Members to complete one online safety training module per month.
- 100% of the office and 75% of the shop have completed one module per month for the past 3 months.
- 25% of Shop Team Members do not see the need for the monthly online safety training.

Extent

Charles F. Kettering said, "A problem well stated is a problem half solved." It was this mindset that enabled him to come from a meager childhood and become one of this country's greatest engineers. In the A3 Problem-Solving process, a well-stated problem underscores the significant ways in which the current situation is different from the standard, expectation, or norm. To truly understand the problem, you must break it down into specific characteristics. These characteristics become the extent of the problem. To determine the extent of the problem, you must ask yourself the following:

- When is the problem happening (time of day, days of the week, etc.)?
- How often does it happen (every hour, every day, every week, every month, etc.)?
- Where is the problem (front, back, top, bottom, process #4, night shift, etc.)?
- How long has it been a problem (month, day, and year if possible)?
- What is the problem doing (staying the same, getting better, or getting worse)?
- What is affected by the problem (people, processes, departments, etc.)?
- What types of occurrences are being experienced (scratches, dents, missing parts, etc.)?

When your initial sense of the problem seems large and vague, you must break it down to get a clear precise picture of the problem. Figure 3.6 shows how defining the extent helps narrow the focus of the problem to specific characteristics. By asking these questions, you should be able to narrow your problem-solving focus to specific characteristics.

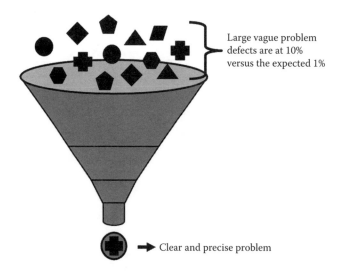

Large vague problem defects are at 10% versus the expected 1%

Clear and precise problem

Figure 3.6   Narrow your problem-solving focus.

In the tank weld scenario, we identified the standard, current situation, and discrepancy. Now we need to determine the extent of the problem by identifying all the relevant characteristics.

- *Standard:* The tank weld Team Member should only take 15 minutes to tack weld the tank together, 1 hour to weld the tank, and 20 minutes to grind all welds.
- *Current Situation:* The Team Member on the tank welding process takes 30 minutes to tack weld the tank together, 1 hour to weld the tank, and 20 minutes to grind all welds.
- *Discrepancy:* It is taking the Team Member 15 minutes longer to complete the process.
- *Extent:*
  - When? Only on "D" shift
  - How often? Every time the "D" shift works
  - Where? Process #3
  - How long? Since January 15
  - What is it doing? Staying the same
  - What is affected? Following processes and the customer
  - What types of occurrences? – Excess time tack welding

To improve the effectiveness of a problem statement, it is sometimes necessary to incorporate these characteristics into the standard, current situation, and discrepancy. The following example incorporates key characteristics into the problem statement to improve clarity.

Tank Weld Scenario

- *Standard:* Tank weld Team Members should take no more than 15 minutes to tack weld the tank together, 1 hour to weld the tank, and 20 minutes to grind all welds.
- *Current Situation:* The "D" shift tank weld operator on process #3 is taking 30 minutes to tack weld the tank together, 1 hour to weld the tank, and 20 minutes to grind all welds.
- *Discrepancy:* The "D" shift weld operator on process #3 takes 15 minutes longer to tack weld.
- *Extent:* The problem has not changed since January 15.

Once you have a precise description of the problem, it is important to find the Point Of Cause (POC) of the problem before continuing. In some situations, you will become aware of the problem at the POC. In this situation, tracking back through the process to find the POC will not be necessary.

Example

A paint operator is spraying parts and notices that the paint is clumping on the part. In this situation, the operator has identified the problem at its point of cause. In other situations where the problem is not identified at the point of cause, you will need to determine the POC before conducting 5-Why analysis.

When the POC is not obvious, you must "track back" to find the POC. The track back process starts where you first identified the problem and consists of walking back through the process steps. You must stop at each point in the process and look at all the product in the process to see if the defect or problem is present. If it is, then you must proceed back to the next process and look to see if the defect or problem is present. This process continues until you get to a point in the process where the defect or problem does not exist. Logic dictates that the operation following the process that you are currently in is where the defect or problem originated. Figure 3.7 shows the track back process. This is the point at which you will eventually begin your 5-Why analysis.

Rationale

Before you can write the rationale, you must evaluate each of your problems to determine which one needs your immediate attention. The evaluation process gives you a better understanding of how each problem fits into the needs of the company, area, or department.

In situations where you are faced with multiple problems, it is wise to conduct the evaluation process prior to spending too much time identifying the extent

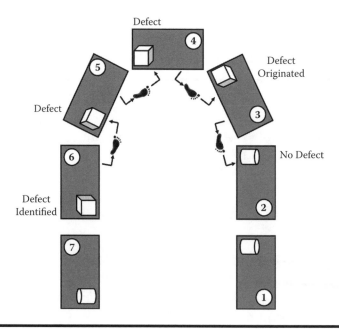

Figure 3.7 Track back process.

and point of cause for any particular problem. It is important, as a problem solver, that you select the right problem at the right time. If you spend time trying to determine the extent and point of cause for several problems, you will be wasting time on the problems that are not selected during the evaluation process. This is typically more of an issue the further you are from the actual process.

The goal of the evaluation process is to determine the need to address a problem now. Prioritizing the problem in the context of other problems occurring at the same time helps you communicate your rationale to others.

When evaluating problems, you should ask what is the

- Importance? How does this problem fit with management objectives related to cost, quality, safety, productivity, delivery, and personnel?
- Urgency? How quickly should this problem be addressed to prevent it from creating other problems?
- Tendency? Based on what you have observed and the data available, what will happen regarding this problem if nothing is done? Will it get worse if left alone? (High) Will it stay the same? (Medium) Will it get better on its own? (Low)

Example

The company is scheduled to provide a new customer with product beginning the first week of January, provided the company meets ISO certification requirements. The new customer will increase sales by 20%. The ISO certification audit is scheduled for October 15 of this year.

- *Problem:* All level 2 documents are to be completed by July 31 of this year. As of July 31, 75% of level 2 documents are not complete.
- *Rationale:*
  - Importance. ISO certification is a high priority for management, as the new customer requires that suppliers be ISO certified prior to submitting product orders.
  - Urgency. This problem must be addressed immediately in order to meet the new customer's ISO certification requirements.
  - Tendency. If no action is taken, documentation will continue to fall behind schedule and the company will not make the certification deadline.

In a situation where there is more than one problem, it may be necessary to highlight why you selected one particular problem over another. Creating a simple grid can simplify the process and make it easier for others to understand why you selected one problem over another. Table 3.2 evaluates the importance, urgency, and tendency of several problems.

During the evaluation process, importance, urgency, and tendency should be rated High, Medium, or Low in order to understand how a problem or problems

Table 3.2　Problem Evaluation Table

Paint Problem	Importance	Urgency	Tendency
Paint seeds 20%	H	H	L
Paint scratches 17%	H	H	H
Thin paint 7%	M	M	L
Paint runs 4%	M	L	L

relate to management objectives. All these factors should be considered when several problems arise at the same time and priorities should be determined.

When developing the rationale for your problem, keep in mind that you are explaining to your supervisors and management why you picked up this problem and why it is important for them to assist you in solving this problem.

Chapter 4

Target

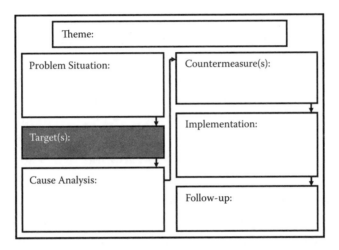

Figure 4.1   A3 problem report. The target is the third block of the A3 and is completed after the problem situation block of the A3.

The Difference between Targets and Goals

After defining the problem situation, it is essential that the problem solver set target(s). At this stage of the process, if targets are not set, procrastination can set in and other things may take priority.

Most companies that I visit do not make a distinction between a target and a goal. In A3 Problem Solving, goals are set by management and represent management's direction for the organization. Some examples of goals might be the following:

- Productivity: Increase sales by 20%.
- Cost: Reduce cost 10%.
- Quality: Improve first-pass yield by 5%.
- Safety: Reduce near misses by 90%.
- Maintenance: Reduce machine downtime by 15%.
- Personnel issues: Increase perfect attendance by 25%.

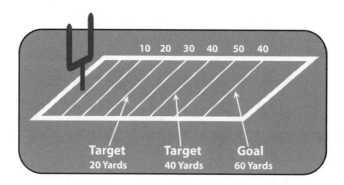

Figure 4.2   Field goal targets.

Driving this information to shop leadership and process Team Members makes it easier for management to achieve their goals and objectives. One way that I've seen this done was through mandatory attendance at monthly and quarterly company meetings. One day per month, the production requirement would be reduced so that all Team Members could be in attendance at a monthly meeting. During the meeting, information on market share, safety, quality, delivery, and even planned Team Member events that are designed to improve morale were shared.

Without this type of information, many organizations will struggle to achieve their objectives because there is no shared agenda. Don't get me wrong; I'm not suggesting that stopping production to have meetings will work for everyone. However, I do suggest that companies find ways to share this type of information on a regular basis to ensure that everyone understands the company's current position in the market and what needs to be done to guarantee the future success of the company.

In a Lean organization or an organization making a Lean transformation, problem solvers are responsible for setting targets to support management's goals. Targets represent incremental movement toward management's goals.

Let's say that you are the special team's coach of your favorite football team. You want your kicker to be able to make a 60-yard field goal 90% of the time while under pressure from the opposing team. To reach this goal, the coach sets incrementally longer targets for the kicker to achieve until the kicker is able to make 60-yard field goal kicks 90% of the time while under pressure. Figure 4.2 represents the targets in relation to the overall goal.

Structure of an Effective Target

A target should be specific as to what you are trying to accomplish as it relates to the problem and all the characteristics identified. To be specific, you need to know what change you expect to see and how you will measure the change. By taking the time to break down the problem and track it back to the POC, it is easier to be specific.

Targets must also be achievable and realistic. This means setting dates that are reasonable in regard to what needs to be accomplished and what constraints apply to the situation. Constraints can be deadlines established by management, customer requirements, resources, or other requirements.

To meet these criteria, a good target statement will contain four basic components:

1. A verb that describes the action you want to take, such as increase, decrease, reduce, or eliminate. Referred to as the "Do What" part of the target in A3 Problem Solving.
2. A brief descriptive statement of the problem based on the characteristics identified in the extent. This statement becomes the "To What" portion of the target.
3. A specific measurement of what you want to achieve based on the standard identified in the problem situation. This is known as the "How Much" part of the target.
4. A specific time frame (month, day, and year) for achieving the result based on the constraints of the situation. This date is the "By When" and completes the target statement.

Example

Do what: Reduce
To what: Outlet Box scrap on molds 13, 15, and 17 due to missing tabs
How much: From 13% to 0.5% or less
By when: August 12, 2009

It is important that you consider your position in the organization and your ability to affect change as it relates to the problem that you identified. If you are unable to drive the needed change at your level in the organization, it is important that you elevate the problem to the appropriate level. Keep in mind that you should not just bring the problem but you must also know why the problem exists and have ideas for resolving the problem. The A3 is the perfect tool for elevating the problem to the appropriate level.

In addition to a target statement, you may also want to include a line graph to visually display your target. Figure 4.3 illustrates an Outlet Box scrap target graph.

Target Statement Exercise

Read the following examples and rate each target statement using B for best, G for good, or N for needs improvement. For those needing improvement, state

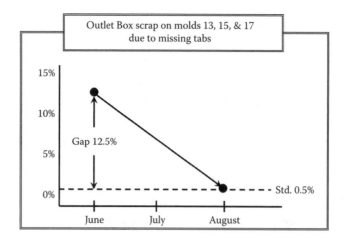

Figure 4.3   Outlet Box scrap target graph.

what should be done to make the target statement more effective. See Appendix E for answers to this exercise.

Example 4.1

A. _____ Increase Safety Committee attendance by 10% before March 25, 2010.
B. _____ Increase Safety Committee attendance of Shop Team Members by 10% before March 25, 2010.

Example 4.2

A. _____ Reduce the time it takes to tap all left side door hinge holes from 15 minutes to 10 minutes by the end of the week.
B. _____ Reduce the time it takes to tap all left side door hinge holes by no later than August 30, 2010.

Example 4.3

A. _____ Eliminate downtime on sewing line #4 by no later than November 7, 2010.
B. _____ Eliminate downtime on sewing line #4 due to broken needles by November 7, 2010.

Example 4.4

A. _____ Improve turnaround time on book binding from 2 weeks to 3 days by April 30, 2010.
B. _____ Reduce book binding turnaround time from 2 weeks to 3 days by April 30, 2010.

Example 4.5

    A. _____ Increase Team Member morale rating on the 2012 opinion survey.

    B. _____ Improve Team Member morale rating from 80% to 95% on the 2012 opinion survey.

Example 4.6

    A. _____ Move to #95 on the list of best places to work in the state by 2010.

    B. _____ Improve Lotta-Lift's ranking on the list of best places to work in the state from #111 to #95 by May 10, 2010.

Chapter 5

Theme

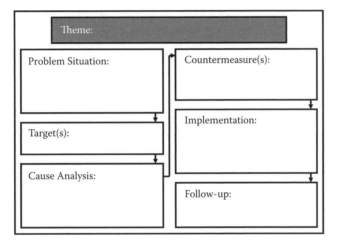

Figure 5.1   The theme is at the top center of the A3 and is completed after the target has been determined.

General Theme Categories

In a Lean production environment, there are several major management themes, which include the following:

- Cost
- Quality
- Safety
- Productivity
- Delivery
- Profit
- Maintenance
- Personnel issues

Tying your problem to one or more of these general themes will influence decisions that you and others make throughout the Problem-Solving process and how the problem is perceived by management.

Creating the Theme

It seems obvious that the theme block, being at the head of the A3 form, would be the first part of the A3 to be completed. But in this case, logic trumps the obvious. The theme provides the reader with the focus of the A3. However, at the beginning of the process, your sense of the problem is probably vague.

Deciding on a theme at this point would not provide the reader with any insight into the true nature of the problem. For that reason, it is important that the problem solver wait until the problem situation is defined and the target has been set.

The theme must capture the essence of what you are trying to achieve and should never exceed one sentence in length. It should capture the attention of the reader by specifying the characteristics of the problem. The theme comes directly from the target statement set by the problem solver.

In most A3 reports that I've read outside of Toyota, the owner usually manages to create a concise theme. However, these themes are usually so concise that the reader does not really understand what the problem solver intends to accomplish. Think about the following examples:

1. Reduce outlet box scrap.
2. Reduce outlet box scrap on molds 13, 15, and 17 due to missing tabs.

In both cases, the reader knows that outlet box scrap is the focus of the problem and what the writer wishes to address. However, in the second example, the writer communicates the focus much more clearly by including which molds and the actual defect resulting in the scrap.

Having already created a clear target statement makes it much easier to state the theme. The most effective and descriptive method for stating the theme is to use the "Do What" and "To What" portions of your target statement, providing that you did a good job breaking down the problem and including the characteristics in the target statement.

Theme Statement Exercise

Identify the "Do What" and "To What" in each of the target statements that follow. Write a theme for each of the target statements. See Appendix F for answers to this exercise.

Example 5.1

- Target: Increase Safety Committee attendance of Shop Team Members from 30% to 50% before March 25, 2010.
- Theme:

Example 5.2

- Target: Reduce the time it takes to tap left side door hinge holes from 15 minutes to 10 minutes by August 30, 2010.
- Theme:

Example 5.3

- Target: Eliminate downtime on sewing line #4 due to broken needles by November 7, 2010.
- Theme:

Example 5.4

- Target: Reduce book binding turnaround time from 2 weeks to 3 days by April 30, 2010.
- Theme:

Example 5.5

- Target: Increase 2012 Team Member morale opinion survey rating from 80% to 95% by March 24, 2012.
- Theme:

Example 5.6

- Target: Improve Lotta-Lift's ranking on the list of best places to work in the state from #111 to #95 by May 10, 2010.
- Theme:

Chapter 6

Loud-&-Clear Speaker

Practice Case Study

May 25, 2010: You are the Team Leader on the 8- to 10-inch speaker assembly line at Loud-&-Clear Speaker located in El Paso, Texas. The speakers produced on your line have a specially designed diaphragm made of a clear plastic-like material that is more effective at reproducing high-quality sound. This technology is currently only used in the production of Loud-&-Clear's top-of-the-line home theater sound systems.

Loud-&-Clear Speaker just engineered a series of high-end trunk speakers that will be produced on the same line as the 8- to 10-inch speakers using the same technology as the home theater systems. Production is scheduled to begin June 15, making the product available for sale in time for the July 4th holiday.

As a Team Leader, you are responsible for training employees using Job Instruction, responding to Andon calls originating from your line, assisting Team Members, relieving them for emergency breaks, and for solving problems that affect your line. In short, you are responsible for keeping the line running and for producing quality products in a timely manner.

You just returned to work after a week-long vacation at White Sands, New Mexico, where you spent time improving your snowboarding skills by practicing on the white gypsum sand dunes. Shortly after lunch, you stopped by the sound test station near the end of the assembly line where the 8- to 10-inch speakers are visually inspected and then tested for clarity of sound. You looked at your department board and noticed that quality defects are up significantly on the 8- to 10-inch speaker line since Monday of last week.

To verify that your initial perception that quality defects have increased significantly is accurate, you compare the standard to the total number of defects on the quality checksheet at the sound test station. There were 1,000 units produced, and 100 of those units had defects. This makes your current defect rate 10%.

As the Team Leader of the 8- to 10-inch speaker line, you know that the quality standard for your line is 1% or less. When the current situation of 10% is compared to the standard of 1%, you can see that the quality defect rate is 9% above the standard.

Stop: Using the above information, write a statement of the background in the blank Problem Situation block provided at the end of this chapter (Figure 6.6). Once complete, return to the section of the text that explains the Standard.

The following is a list of speaker parts and their functions, and Figure 6.1 shows the location of each of the speaker parts:

- Suspension: holds the diaphragm to the basket while allowing flexibility to move up and down
- Dust Cap: keeps dust and debris from interfering with the movement of the coil as it moves up and down
- Diaphragm: transforms the electrical sound waves into the words and music that you hear when you listen to the radio or television by interacting with the air around the diaphragm
- Basket: provides the structure and connecting points for all of the speaker components
- Spider: allows the cone to return to its neutral position once the input is terminated
- Voice Coil: raises and lowers the diaphragm when power is applied
- Magnet: attracts or repels the voice coil to make the vibrations that create sound

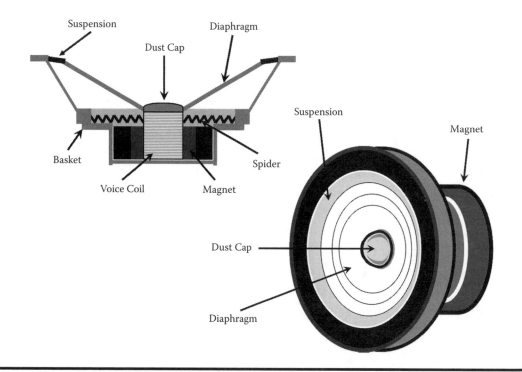

Figure 6.1   Speaker part names and locations.

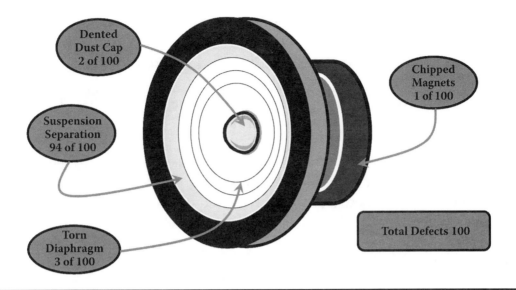

Figure 6.2   Defect types.

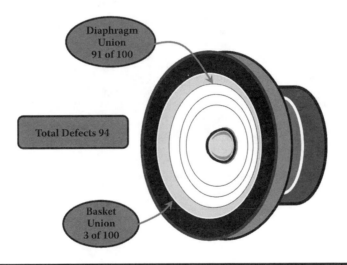

Figure 6.3   Location of selected defect.

As the Team Leader for the 8- to 10-inch speaker line, you decided to break down the types of quality defects; Figure 6.2 shows each defect type. You identified four different **"types"** of defects that make up the total of 100 defects on the 8- to 10-inch speaker line:

- 94 suspension separation defects
- 3 torn diaphragms
- 2 dented dust caps
- 1 chipped magnet

There are two places "where" suspension separation can occur. You decide to further break down the 94 suspension separations into the area between the basket and the area where the suspension attaches to the diaphragm. Figure 6.3 shows where the selected defect is located.

The suspension separation defects break down accordingly:

- 91 where the diaphragm and the suspension meet
- 3 where the suspension attaches to the basket

Now you decide to look at "when" the 91 suspension separations between the diaphragm and the suspension occur by breaking it down between Blue and Red shifts; Figure 6.4 shows how many of the selected defects occur on each shift.

The defects break down accordingly between the shifts:

- 90 on Blue shift
- 1 on Red shift

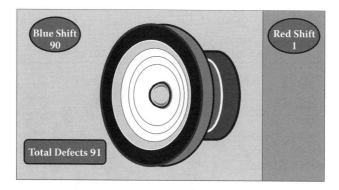

Figure 6.4   Number of defects by shift.

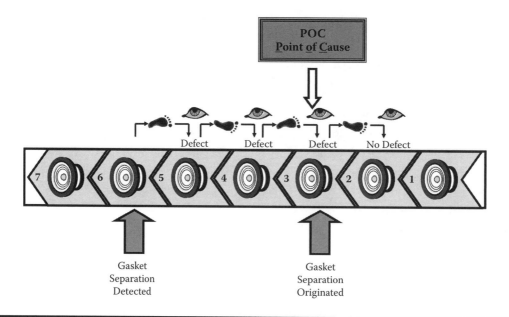

Figure 6.5   Track back process.

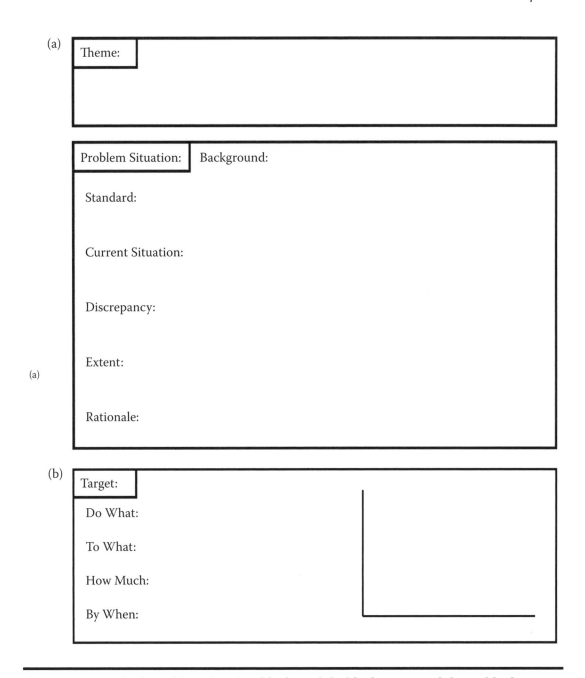

Figure 6.6 **(a) Blank problem situation block, and (b) blank target and theme block.**

You still feel like you are missing some information about where the defects are "originating." To find this out, you begin to track back, beginning with process #6, the sound test station where the problem was detected; Figure 6.5 is a visual representation of the track back process.

At process #5, you inspect the product at the process. This includes reviewing any inventory or batch product. You see that the problem is evident at process #5. You continue tracking back to process #4 and look at the product. The problem still exists at process #4. After repeating this process several times, you find no evidence of the problem at process #2. When you no longer see the problem, you realize that the POC (Point Of Cause) is process #3 where you last saw the problem.

Write a clear Problem Situation, Target, and Theme using the information in the Loud-&-Clear case (Figures 6.6a and b are blank Theme, Problem Situation, and Target). See Appendix G for answers to this exercise.

Chapter 7

Cause Analysis

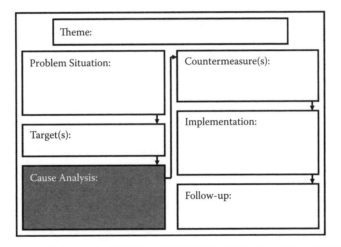

Figure 7.1 Cause analysis is the fourth step in the PLAN phase in A3 Problem Solving.

5-Why

Cause Analysis (Figure 7.1) is the fourth step in the PLAN phase of the A3 Problem-Solving process. Once the problem is identified, there must be a clear understanding of why it occurred before taking any action to correct the situation. The objective of Cause Analysis is to identify the root cause(s) of the problem so that it (they) can be eliminated or minimized.

Another term for cause analysis is "5-Why." This is a technique used to ensure that you ask why enough times to uncover the real root cause(s) of a problem. In many of the facilities I have visited, 5-Why means asking for five different reasons as to why the problem occurred, none of which proceed logically to the root cause. The true nature of 5-Why is to ask why in a manner that leads you from the problem to the root cause. A critical part of the 5-Why process is to investigate the causes at each level to find out which ones are facts and to let the facts lead you to the root cause.

Shinichi Imaeda, my Japanese coordinator at Toyota Motor Manufacturing Kentucky (TMMK), explained that the number "5" is not so much a strict rule but rather a guideline used to ensure that, as you investigate a problem, you ask why enough times to reach the real root cause.

The other reason the Japanese use the number "5" is because many odd numbers in Japan are considered lucky; and when it comes to solving problems, a little good luck is not a bad thing. It could have been called "3-Why" but it was felt that by only asking why three times, you would not be digging deep enough to get to the real root cause. Imaeda also explained there will sometimes be fewer than 5-Whys and sometimes more than 5-Whys, depending on the situation.

To be effective at solving problems, you must be able to build a succession of cause-and-effect relationships that direct you from the problem to the root cause. The process of creating these cause-and-effect relationships known as 5-Why must be based on fact at each step, not assumption. This point as well as the importance of breaking down the problem in the Problem Situation has been an eye-opener for more of my students than anything else. Many of my students were under the impression that a 5-Why chain was a series of "I thinks" rather than "I knows."

Problem: Sealant gaps on 15% of glass inserts.

(Why?)
"I think" it is because of this.
(Why?)
"I think" it is because of this.
(Why?)

And so it goes until they "think" they have reached the root cause.

This process of getting facts is an absolute necessity in the A3 Problem-Solving process. However, this process does not proceed in a straight line from problem to root cause. There may be several detours along the way to identifying the root cause(s).

Figure 7.2 provides a flowchart that can guide you through the process of creating a 5-Why chain based on fact rather than on assumption.

Cause Analysis Process

Start at the "Point Of Cause" by asking yourself, "Based on the facts of the current situation, what could be causing the problem?" The answer to this question can take you in one of two directions. Either you think you know why, or you do not know why.

- If you think you know, check to confirm your belief. Once confirmed, the answer becomes the next link in the cause-and-effect chain.

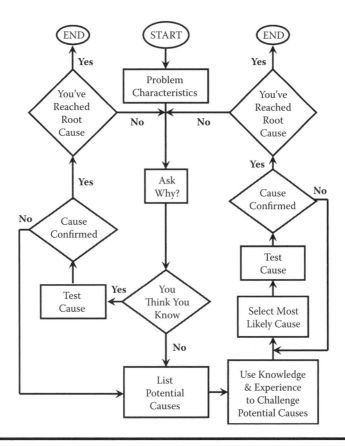

Figure 7.2   Cause analysis 5-Why flowchart.

■ If you do not know the answer, then you must list potential causes that could explain why the problem exists. To generate meaningful potential causes, think in terms of Mankind, Method, Material, Machine, and Environment, otherwise known as the 4Ms and 1E.

Example

You are the third-shift Maintenance Team Leader and it is Sunday night, the beginning of the workweek. You have been called to the packaging line for the third week in a row because the ink jet printer will not print. The last two times the ink jet would not print was because it had dried ink clogging the nozzle. It always seems to happen after the equipment has been shut down over the weekend. From the time you get the call until the printer is functioning normally, it takes about 20 minutes. Figure 7.3 depicts the potential causes of the problem.

Problem: 20 minutes downtime on packing line #2 due to label printer not printing.
  WHY?
The ink nozzle is clogged with dried ink.

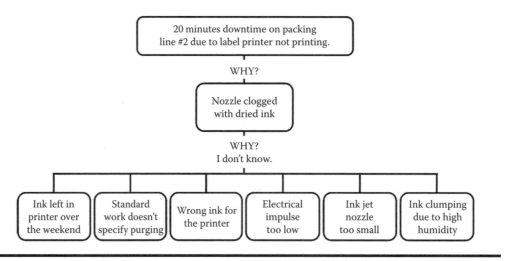

Figure 7.3   List of potential causes.

WHY is the ink dry? I don't know.

How could **mankind** (associate) have contributed to the dry ink clogging the nozzle?
  – Associate could have left ink in the system at the end of the week.
How could the **method** used by the associate contribute to the ink nozzle clogging?
  – Standardized work may not specify purging the ink from the system at the end of the week.
How could the **material** have contributed to the ink nozzle being clogged?
  – The ink may not meet requirements for the ink jet printer.
How could the **machine** contribute to the printer downtime?
  – Electrical impulse used to spray ink is too low, resulting in ink staying in the nozzle.
  – Ink jet nozzle may be too small for the printer, thus not allowing ink to spray properly.
How could the **environment** contribute to the nozzle being clogged?
  – Ink clumping in nozzle due to high humidity.

It is essential to understand that analyzing for root cause is not the same as identifying potential causes. When analyzing for root cause, you build a chain of cause-and-effect relationships leading to the root cause.

When you identify potential causes, you list all the possible causes of a problem. It is important to note that one potential cause does not lead to another potential cause. Each potential cause must be considered independently. You will need to challenge each potential cause using your knowledge and experience. Your knowledge will make it possible for you to make reasonable assumptions about which potential cause is the most likely source of the problem.

Example

Because it is winter and the humidity is much lower than it is in the summer, it is unlikely that humidity could be causing the problem. It is also unlikely that the ink jet nozzle being too small would cause this problem. If the nozzle being too small were the cause, it would happen more frequently.

If the electrical impulse needed to super heat and project the ink from the nozzle were too low, this would happen all the time, not just one time per week.

Because there are no further problems after the ink nozzle is cleaned, it is unlikely that the ink itself has any bearing on the situation.

Based on the information you have on the other potential causes, the two remaining potential causes seem to be the most likely place to start because you are unable to make any reasonable assumptions about either one. You will have to study the standardized work in order to determine if purging the ink jet printer is required. If purging the ink jet printer was specified, you will need to talk with the associate to see if the printer was purged correctly.

Challenging each of the potential causes with your knowledge and experience enables you to save time because you are setting aside the less likely potential causes and starting your investigation with the most likely cause or causes. Figure 7.4 highlights the most likely causes.

Once you have isolated what you feel is the most likely potential cause, confirm the cause by checking to see if it actually occurred. The most likely potential cause based on the available information is that the system was not purged at the end of the shift. You must find a way to confirm the cause. If the cause cannot be confirmed, you will need to check the other potential causes.

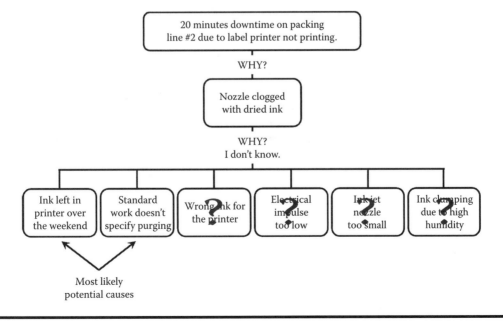

Figure 7.4   Most likely potential causes.

If the cause is confirmed, continue to follow the process to establish a chain of cause-and-effect relationships until you have reached the root cause of the problem.

After reviewing the standardized work, you find that the ink jet printer is supposed to be purged every Friday at the end of second shift. You continue your investigation by asking the associate if the system was purged. You find that the associate is new and did not know the system needed to be purged at the end of the shift. After talking to the Team Leader, you find out that this is the first new associate to be trained in the process in the past 3 years. You also find out that the purge process was overlooked when the company began using Job Instruction Teaching Methods and Training Time Tables. Figure 7.5 reveals the 5-Why causal chain to root cause.

Remember that the potential cause with question marks were set aside, not eliminated, based on your knowledge and experience. The remaining potential cause will need to be checked to ensure that each of the potential causes does not in any way contribute to the problem. For each potential cause, you will have to determine a method to eliminate or affirm it as a cause.

If you were going to check to see if the electrical impulse used to heat the ink and project it out of the nozzle is adequate, you would probably check the manufacturer's specifications and then test the amount of electricity at the

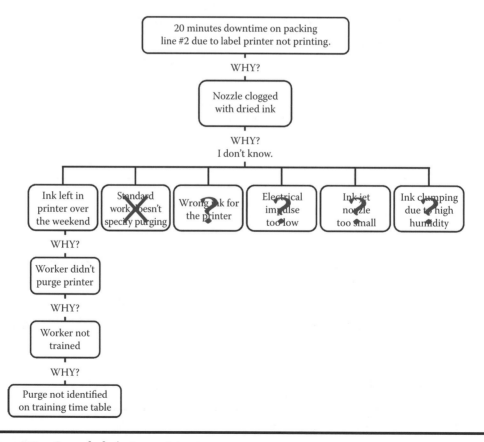

Figure 7.5   Causal chain to root cause.

20 minutes downtime due to packing line #2 label printer not printing
Why?
↳ Printer nozzle clogged
Why?
↳ Dry ink in printer nozzle
Why?
↳ Ink left in printer over weekend dries
Why?
↳ Worker did not purge ink from printer
Why?
↳ Worker not trained on purging process
Why?
Root Cause: ↳ Purge process not identified on training time table

Figure 7.6   A3 report 5-Why chain.

nozzle. The results of the test will tell you if the electrical impulse meets the manufacturer's specifications. You will continue this process until each cause has been checked.

The information obtained from cycling through the cause-and-effect process is used to create your 5-Why causal chain. Figure 7.6 shows what a 5-Why analysis looks like when placed in an A3.

In the quest for the root cause or causes of a problem, you must gradually peel back the layers that lead from the problem to the root cause(s). To be effective, you must gather the facts related to each of the potential causes, making it possible to sift through what is and is not happening in relation to the problem. As you progress to each new level, you eliminate potential causes, thereby allowing you to focus on the ones that lead to the root cause(s).

Stop asking why at the point you feel you have reached the root cause and then evaluate what you feel is the root cause.

Determining Root Cause

There are three attributes you should look for when determining if you have reached the real root cause. By comparing what you think is the root cause with the following attributes, you will be able to tell whether or not you have found the real root cause:

1. The cause is a logical source of the discrepancy.
2. The cause can be addressed directly.
3. The cause is an element for which you can plan countermeasures.

In the example of the downtime on the label printer, the root cause is that the purge process was not identified on the Training Time Table as a training need. Ask the following questions of that root cause:

■ Did you ask why enough times that you have something specific you can countermeasure? Yes, you can add it to the Training Time Table so that it is not overlooked in the future.
■ If you ask why again, will it take you to another problem? Yes, you would be asking how the purge process was left off the list of required trainings.
■ If you implement a countermeasure that addresses the root cause, will it prevent the problem from recurring? Yes, by adding the purge process to the Training Time Table, the Team Leader will not overlook the training next time a new person is trained on the process.

If you are able to answer yes to all three questions, you will have a higher likelihood of successfully achieving your target. If addressing what you believe is the root cause(s) will not allow you to achieve or bring you closer to your target, you will need to continue with the Cause Analysis process until you identify the real root cause(s).

One of the benefits of the A3 is that the information can be shared easily with other departments and even sister companies. In the years that I have been teaching A3 Problem Solving, I often find that leaders in different departments experience problems that stem from similar root causes. Many of these problems could be avoided by sharing the findings of A3 Problem Solving throughout the organization, thus allowing leaders to learn from one another.

Therefore Test

Before moving on to the countermeasure portion of the A3 Problem-Solving process, you must test the logic of your Cause Analysis. To do this you will use the word "therefore" in place of the word "why" and read the chain of events in reverse. This is referred to as the "Therefore Test" and helps identify information that might be missing from your analysis, making it difficult for someone with little or no knowledge of the problem to understand.

Start by reading aloud the statement of the root cause followed by the word "therefore." Then read the statement previous to the root cause and say "therefore." Continue this process until you have read the statement of the problem at the beginning of the 5-Why causal chain.

At any point in the process, if you are not able to insert the word "therefore" and make sense of the analysis, you have a gap in your logic. If there are gaps, you will need to fill in the gaps with additional facts. Figure 7.7 shows how using the "Therefore test" can help you detect an incomplete 5-Why chain.

Read the following cause-and-effect 5-Why chain to see if there are any gaps: To those with maintenance experience, the following analysis may seem complete because they can make intuitive leaps based on their experience and knowledge that close the gaps. However, to those with little or no maintenance experience, it may be a little difficult to make the leap from the associate not purging the ink

20 minutes downtime due to packing line #2 label printer not printing

Therefore
└── Printer nozzle clogged

Therefore
└── Worker did not purge ink from printer

Therefore
Root Cause: └── Purge process not identified on training time table

Figure 7.7 Incomplete 5-Why chain.

20 minutes downtime due to packing line #2 label printer not printing
Therefore
└→ Printer nozzle clogged
Therefore
└→ Dry ink in printer nozzle
Therefore
└→ Ink left in printer over weekend dries
Therefore
└→ Worker did not purge ink from printer
Therefore
└→ Worker not trained on purging process
Therefore
Root Cause: └→ Purge process not identified on training time table

Figure 7.8 Complete 5-Why chain.

from the printer to the printer nozzle being clogged. Figure 7.8 is based on the same situation but with additional information that makes it flow more logically.

As with most rules, there is an exception to the 5-Why process. There will be times when you will not be able to confirm potential causes due to the inability to get accurate information. You will begin the process just as described earlier. However, at some point in the process, you may not be able to answer why something is happening.

This may be due to the inability to see what is going on. If you are having a problem with painted parts that are coming out of an oven, you may need to brainstorm potential causes regarding how the oven contributes to the problem. However, because you cannot go inside the oven while it is running, you may need to brainstorm a broad spectrum of countermeasures and stagger the implementation to determine what actually caused the problem.

This process is more time consuming than cause-and-effect analysis. Cause-and-effect analysis is like firing a scoped rifle that has been sighted in to a hit a bull's-eye at 100 yards. You are very likely to hit the target dead center provided your scope is dialed in properly. Conversely, the trial-and-error approach is much more like firing a shotgun at the same target, where you will have lots of holes in the target, but only a few will actually hit the bull's-eye.

Now that you have identified the problem and analyzed to root cause, you will need to generate possible countermeasures and select the countermeasure(s) that will allow you to achieve your stated targets.

Production 5-Why Exercise

Read the following examples and number them in the correct sequence to create a logical 5-Why chain leading from the problem to the root cause. To ensure that you have each statement in the correct order, use the "Therefore" test to check your logic. See Appendix H for answers to this exercise.

Example 7.1

Problem: 25% of fabricated fireproof storage boxes do not fit in home location.

_____ Fireproof storage boxes are ¼ inch too tall, wide, and deep.
_____ Associate #3's tape measure is off by ¼ inch.
_____ Associate #3's tape measure is not documented by Quality department.
_____ Associate #3's tape measure is not calibrated.
_____ Associate #3 is using his personal tape measure.

Example 7.2

Problem: 10% of runners on plastic molded parts do not fall onto regrind chute.

_____ Robotic arm not aligned with regrind chute.
_____ Regrind chute is too narrow.
_____ Runner sways back and forth before dropping.

Example 7.3

Problem: 30% of refrigerator water filter cartridges strip out on second shift.

_____ Associate thinks cartridges are too loose.
_____ Torque set too high on cartridge air tool.
_____ Associate does not understand torque requirements for part.
_____ Associate increases air pressure.
_____ Torque specification not specified on Job Breakdown sheet.
_____ Associate not provided specifications during Job Instruction.

Example 7.4

Problem: 18% of inserts damaged when pressed into part.

_____ Neck on press blocks assembler's view.
_____ Insert not aligned flush with part opening.
_____ Assembler cannot see insert in relation to part opening.

Office 5-Why Exercise

Read the following examples and number them in the correct sequence to create a logical 5-Why chain leading from the problem to the root cause. To ensure that you have each statement in the correct order, use the "Therefore Test" to check your logic. See Appendix I for answers to this exercise.

Example 7.5

Problem: Picnic tables have trash stuffed into gaps between boards.

_____ Associates do not want to walk 150 feet to dumpster.
_____ Associates do not put trash in dumpsters.
_____ Associates think it takes too much of their breaks and lunchtime.

Example 7.6

Problem: Interoffice mail not delivered at 9:00 a.m., thus delaying payroll process.

_____ Security was asked to transport associate to Urgent Treatment.
_____ Security did not pick up mail until 8:30 a.m.
_____ Night shift supervisor did not want to take the Team Member to Urgent Treatment.
_____ Mail not sorted until 9:00 a.m. by office staff.
_____ Supervisor did not want to stay late to complete production reports.

Example 7.7

Problem: Network copier jammed.

_____ Paper curls around roller.
_____ Paper is too large.
_____ Paper is feeding at the wrong angle.
_____ Too many pages are feeding at the same time.

Example 7.8

Problem: Supply room ran out of blue ink pens.

_____ Supply room replacement not specified when office supply staff went on vacation.
_____ Supply room staff did not notify office manager.
_____ Office manager did not know that blue ink pens were needed.
_____ Purchasing did not receive a purchase order for blue ink pens.
_____ Purchasing did not order blue ink pens.

Chapter 8

Countermeasures

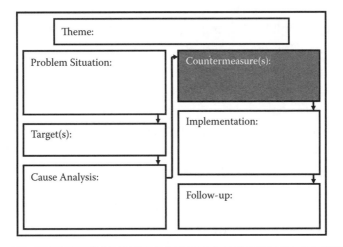

Figure 8.1   The countermeasures section is the last part of the PLAN phase in A3 Problem Solving.

Selecting the "Right" Countermeasures

The Countermeasures section (see Figure 8.1) is the last part of the PLAN phase in A3 Problem Solving. There are four steps to selecting the right countermeasures to help you achieve your targets:

1. Brainstorming countermeasures
2. Narrowing countermeasures list
3. Evaluating remaining countermeasures
4. Selecting the best countermeasure(s)

I will break down each step to provide you with the details needed to ensure that, as you apply the A3 Problem-Solving process, you are able to effectively address the cause(s) of the problem.

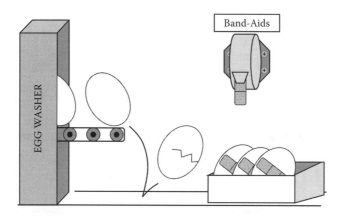

Figure 8.2 Real root cause is neglected.

I begin this section by clarifying the term "countermeasure" as it relates to the thinking in A3 Problem Solving. Many see countermeasures as the solution to the problem. However, in A3 Problem Solving, there is a subtle difference between a countermeasure and a solution. A countermeasure is not the solution but an action taken to implement a counterreaction that helps you achieve your target. Think of it like this: You have an allergy problem that is making your eyes red and irritated, so you take an allergy medication. Taking the medication is an action; the medication blocks histamines that cause your eyes to become red and irritated, thus solving your problem of red and irritated eyes.

There are also two types of countermeasures: long term and short term. Long-term countermeasures deal with the root cause(s) allowing you to achieve your target and prevent recurrence of the problem. Short-term countermeasures will usually affect some level within the 5-Why cause-and-effect chain other than the root cause(s) and will either bring temporary relief from your problem or bring you closer to your target. A short-term countermeasure in many cases is like applying a Band-Aid. Once the wound heals, you take it off. Figure 8.2 represents those situations where a problem is identified and a short-term countermeasure is put in place and the real root cause is neglected.

There are times when short-term countermeasures are more practical than taking an action that will address the root cause. Sometimes in the workplace things change as a result of customer demands or because a new product or service is offered. These changes may, in fact, eliminate the root cause of the problem, making a long-term countermeasure unnecessary.

Brainstorming Countermeasures

Although you may start the A3 Problem-Solving process by yourself, it will eventually require you to include people from your area or other departments, especially as you try to find the right mix of countermeasures to get the results you expect.

There are two ways to approach brainstorming countermeasures. The first is to gather a group of people together to represent your area, other areas that could be impacted by this problem, and areas that are not affected by the problem to brainstorm possible countermeasures. The second is to have that same group of people start the brainstorming process in isolation.

Gathering everyone together at the beginning makes it easier for the facilitator to explain the rules and process that will be used during the session. The drawback to this approach is that you may not get full utilization of a person's creativity. Many people, including myself, need time for the idea or situation to mull around in our heads before we can effectively participate in this type of activity. My best ideas come when I am taking a shower or walking my dogs. Walking my dogs has been the source of many great ideas for my daily work. There are other people who may be asked to participate in the brainstorming process because of their knowledge of the process. Keep in mind that although they may be very knowledgeable, they may also be shy or unwilling to expose themselves to ridicule if one of their ideas is seen as ridiculous by members of the group.

After gathering everyone together, start by reviewing flip charts of the Theme, Problem Situation, Target, and 5-Why chain so that everyone has the same understanding. Once everyone is on the same track, give the group a couple of minutes to think about the situation before beginning the brainstorming process. Providing a few minutes for reflection before the brainstorming process is like priming the carburetor on your lawnmower before you pull the starter cord. Priming the carburetor pushes fuel into the carburetor so that you do not have to pull the cord as many times to get the engine to start.

The second method for beginning a brainstorming session is to distribute a copy of the A3 that has the Theme, Problem Situation, Target, and Cause Analysis sections completed, along with instructions for generating possible countermeasures for the problem. The instructions should also include a "no later than" date for each person to return their list of possible countermeasures. This method provides the necessary time for those who need to mull things over before participating in this type of activity. It also protects those who are shy about sharing ideas openly with the group.

Once everyone has provided a list, you should schedule a meeting to gather everyone together to complete the countermeasure process. In addition to the flip charts that were discussed in the previous method, you will need to post flip charts with all of the possible countermeasures that were submitted.

Regardless of which method you use, the procedure is the same once the group meets. During the meeting you should review the A3 process again and all possible countermeasures that were submitted. Ask the participants to consider possible actions that will neutralize each statement in the 5-Why chain or fish-bone diagram. Asking group members to think about each line in the 5-Why chain as they brainstorm will ensure a good mix of both short-term and long-term countermeasures.

Using a process called "round robin," you can quickly facilitate getting ideas for the possible countermeasures. Start at one end of the group and progress around the room asking for additional input. Continue going around the room until all ideas have been exhausted.

All rules normally associated with brainstorming apply: Stress quantity over quality, do not judge ideas, list ideas as swiftly and accurately as you can, and allow piggybacking (building on someone else's idea to create a new idea).

There was one idea that was submitted during a countermeasure brainstorming session for a paint problem. One of the problem's causes was that the paint was getting thick because the paint was too cold. During the session, one of the participants blurted out, "Use hot pink paint!" Although the rules of brainstorming prohibit judging ideas, the group gave a chuckle. Piggybacking on hot pink paint, someone came up with wrapping heat trace around the paint lines that feed the paint booth to keep the cold temperatures from affecting the viscosity of the paint.

The key to facilitating an A3 brainstorming session is not to let it drag on too long. Once the group has exhausted its creativity, do not continue to prod them for more ideas. The perception will be that you have a countermeasure in mind that they have not identified. During a brainstorming session, you can generate so many ideas that the group can feel overwhelmed, so you need to narrow the list to a manageable number. Figure 8.3 shows a list of countermeasures that were brainstormed for paint spits on frames leaving the paint department on Tuesdays and Thursdays.

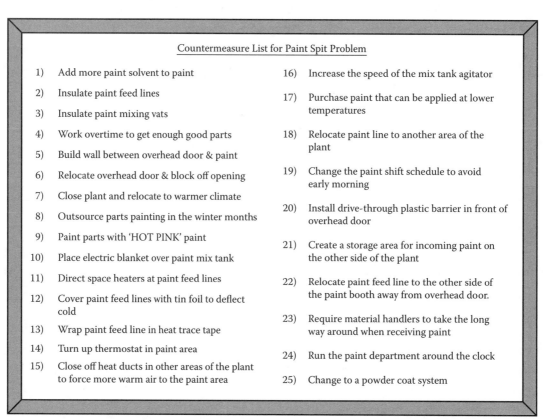

Countermeasure List for Paint Spit Problem

1) Add more paint solvent to paint
2) Insulate paint feed lines
3) Insulate paint mixing vats
4) Work overtime to get enough good parts
5) Build wall between overhead door & paint
6) Relocate overhead door & block off opening
7) Close plant and relocate to warmer climate
8) Outsource parts painting in the winter months
9) Paint parts with 'HOT PINK' paint
10) Place electric blanket over paint mix tank
11) Direct space heaters at paint feed lines
12) Cover paint feed lines with tin foil to deflect cold
13) Wrap paint feed line in heat trace tape
14) Turn up thermostat in paint area
15) Close off heat ducts in other areas of the plant to force more warm air to the paint area
16) Increase the speed of the mix tank agitator
17) Purchase paint that can be applied at lower temperatures
18) Relocate paint line to another area of the plant
19) Change the paint shift schedule to avoid early morning
20) Install drive-through plastic barrier in front of overhead door
21) Create a storage area for incoming paint on the other side of the plant
22) Relocate paint feed line to the other side of the paint booth away from overhead door.
23) Require material handlers to take the long way around when receiving paint
24) Run the paint department around the clock
25) Change to a powder coat system

Figure 8.3   List of countermeasures.

Narrow Countermeasure List

Through the brainstorming process, it is likely that you will have a sizable list of possible countermeasures, many of which may not be effective or feasible. In addition, some of the countermeasures may impact the other areas of the organization in a negative way. To select the best countermeasure(s) for the situation, it will be necessary to narrow the list to a manageable number of countermeasures that are effective, feasible, and have an overall positive impact on the situation and organization.

The method generally used in A3 Problem Solving to narrow the list of countermeasures is called N over 3 (N/3). The idea is to divide the total number of countermeasures by 3 and let each Team Member place votes next to the countermeasures they feel are the most effective, feasible, and will have the greatest positive impact on the problem based on their experience and the constraints of the situation. Prior to applying N/3 and these criteria, it is helpful if you can gain a consensus from the group about which countermeasures can be eliminated based on common sense. If a countermeasure is obviously unworkable or nonsensical, it should be stricken from the list. N/3 Rules include:

1. Participants are not required to use all their votes.
2. Participants may only use one vote per countermeasure. In other words, if you have six votes, you may not apply all six votes to one countermeasure. Appling all your votes to one countermeasure will skew the results of the narrowing process based on personal feelings versus what the group feels are the best options.

As a member of an A3 Problem-Solving event, you must apply criteria to your thought process as you vote on which countermeasures should be passed on to the next part of the selection process.

As you consider each of the countermeasures listed, ask yourself, "Based on the facts of this problem situation, is this countermeasure likely to be effective, feasible, and impact the situation in a positive manner?"

- Effective: Ask, will this countermeasure help achieve the targets established earlier in the A3 Problem-Solving process?
- Feasible: Ask, is this countermeasure possible given the circumstances and constraints of this situation?
- Impact: Ask, does this countermeasure create problems that might outweigh any potential benefits of the countermeasure?

During this part of the countermeasure process, these three criteria are used in a more general sense to pare down the list. Later in the process, these same criteria will be used to thoroughly evaluate the remaining countermeasures that survive this phase.

Previously I gave the example of "hot pink paint." This is obviously a nonsensical option for solving the problem of the paint being too thick. It, along with any other ideas that are similar in nature, should be eliminated before narrowing down the list of countermeasures.

During the narrowing process, your common sense along with the criteria should allow you to set aside any countermeasures that the group may feel are less effective, less feasible, and may have little or no positive impact on the situation. Keep in mind that if the narrowed-down list does not produce any usable countermeasures, it may be necessary to fall back on some of the countermeasures that were set aside during the narrowing process or to brainstorm additional countermeasures.

Once the list has been narrowed down, it is important that you put the remaining countermeasures through a more stringent evaluation to determine which countermeasure(s) will provide the best results. Figure 8.4 is the list of countermeasures after all nonsensical countermeasures have been deleted and displays the results of the groups N/3 (23/3) voting in which each member of the group was allotted a maximum of 7.6 or 8 votes.

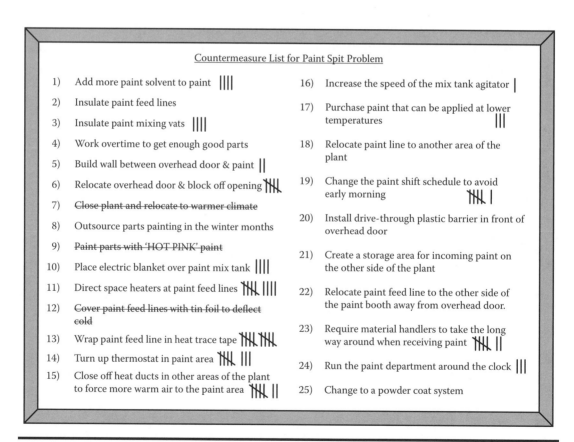

Figure 8.4 **Narrowed-down list of countermeasures.**

Evaluate Remaining Countermeasures

After the list has been narrowed down, the remaining countermeasures must be completely evaluated in order to rank each countermeasure as to its projected effect on the problem. To make an effective projection, you will need to use the same criteria used in the narrowing process but in a more in-depth manner.

Effectiveness

The first thing that should be evaluated is the countermeasure's ability to do two things: (1) bring you closer to your stated target(s) and (2) prevent the recurrence of the problem. Ask yourself:

- Will implementing this countermeasure bring you closer to your target(s)?
- Will implementing this countermeasure achieve your target(s)?
- Will implementing this countermeasure prevent a recurrence of the problem?

Some of the countermeasures may only provide temporary relief of the symptoms. These countermeasures are referred to as short term and play a vital role in helping keep the problem in check. Once the problem has been clearly identified, it may be necessary to contain the problem through the use of short-term countermeasures.

You must also seek long-term countermeasures that address the root cause(s) of the problem in order to keep the problem from reappearing. Often, it is a combination of long-term and short-term countermeasures that provide the best results. Figure 8.5 is the scale used to rate a countermeasure's effectiveness.

To evaluate the effectiveness of a possible countermeasure, use the scale:

- High: High probability the countermeasure will achieve the stated target(s)
- Medium: Moderate probability the countermeasure will achieve the stated target(s)
- Low: Low probability the countermeasure will achieve the stated target(s)

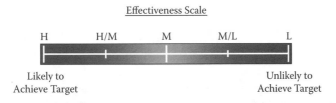

Figure 8.5 Countermeasure effectiveness scale.

Feasibility

Countermeasures must be evaluated to determine how practical the countermeasures are to implement. Figure 8.6 is the scale used to rate a countermeasure's feasibility. Question the feasibility of each countermeasure with regard to:

- Will it cost more than it saves?
- Will it compromise quality?
- Will it compromise safety?
- Will it require additional resources (people, material, time, etc.) to implement?
- Will it be approved?

To evaluate the feasibility of a possible countermeasure use the following scale:

- High: High probability the countermeasure will be approved
- Medium: Moderate probability the countermeasure will be approved
- Low: Low probability the countermeasure will be approved

Impact

Regardless of how effective a countermeasure is or how feasible it will be to implement, it can still have adverse effects on others. Adverse effects could be as simple as asking an associate or another department to complete a checklist to ensure quality or to prevent the omission of critical process steps. Figure 8.7 is the scale used to rate the impact of a countermeasure.

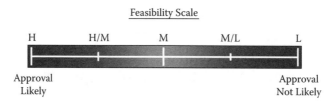

Figure 8.6   Countermeasure feasibility scale.

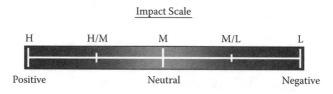

Figure 8.7   Countermeasure impact scale.

You will need to determine in what ways and to what degree the possible countermeasure could affect:

- The associate
- Your job
- The department
- Other processes
- Other departments
- The organization

To evaluate how significant the impact of the countermeasure is, use the following scale:

- High: It is likely to have a positive impact on others.
- Medium: It is likely to have a neutral impact on others.
- Low: It is likely to have a negative impact on others.

Now that the criteria have been defined, each of the countermeasures on the narrowed list should be evaluated.

Figure 8.8 is the 5-Why chain for the paint spit problem, and Figure 8.9 is the plant layout for the problem.

The key to evaluating countermeasures is to evaluate the countermeasures horizontally—not vertically. Evaluate each countermeasure using all three criteria before moving on to the next. By following this process you will be less likely to make a comparative analysis that may end up as a reflection of your preconceived ideas rather than an impartial evaluation of the countermeasures.

The group decided to conduct a more thorough evaluation of the countermeasures that received five or more votes. The seven countermeasures that received

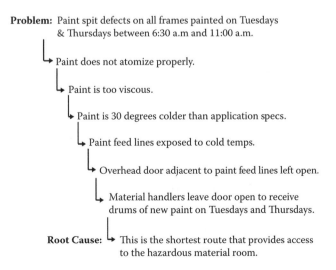

Problem: Paint spit defects on all frames painted on Tuesdays
& Thursdays between 6:30 a.m and 11:00 a.m.

↳ Paint does not atomize properly.

 ↳ Paint is too viscous.

  ↳ Paint is 30 degrees colder than application specs.

   ↳ Paint feed lines exposed to cold temps.

    ↳ Overhead door adjacent to paint feed lines left open.

     ↳ Material handlers leave door open to receive
drums of new paint on Tuesdays and Thursdays.

Root Cause: ↳ This is the shortest route that provides access
to the hazardous material room.

Figure 8.8   Paint spit problem 5-Why chain.

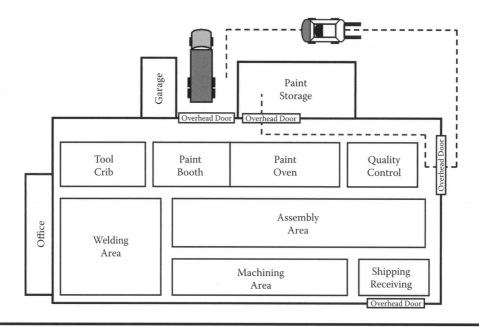

Figure 8.9 Plant layout.

five or more votes are listed below and will be transferred to a countermeasure evaluation table for the next step in the process:

1. Relocate the overhead door and block off the opening.
2. Direct space heaters at paint feed lines.
3. Wrap paint feed lines in heat trace tape.
4. Turn up thermostat in paint area.
5. Close off heat ducts in other areas of the plant to force more warm air to the paint area.
6. Change the paint shift schedule to avoid early morning.
7. Require material handlers to take the long way around when receiving paint.

The following information explains the effectiveness, feasibility, and impact of each of the countermeasures:

1. *Relocate the overhead door and block off the opening.* Removing the overhead door and bricking in the opening would prevent cold air from entering the area and affecting the temperature of the paint. It would also address the root cause of the problem and prevent recurrence of the problem. There would be some expense to move the door and block the opening but it would be offset by the reduction in rework. There would be an additional benefit in that the associates' work area would be warmer.

2. *Direct space heaters at paint feed lines.* Maintenance has space heaters that are used when maintenance has to work outside, so it may be possible to use the heaters without purchasing additional space heaters. The space

heaters would keep the paint in the paint feed lines from getting too cold. The drawback may be that the paint would get too warm, thus creating a different type of quality problem.

3. *Wrap paint feed lines in heat trace tape.* Applying heat trace will allow you to regulate the temperature of the paint in the paint feed lines, allowing the paint to atomize properly. Heat trace is relatively inexpensive and easy to install. It should have a positive impact for all involved if there are no defects to repair.

4. *Turn up thermostat in paint area.* Turning up the heat will increase the temperature in the paint area. However, most of the heat will rise to the ceiling. When the overhead door is opened, the additional heat will escape. Turning up the heat in the area would have minimal effect on the problem and create additional utility costs.

5. *Close off heat ducts in other areas of the plant to force more warm air into the paint area.* Closing off the heat ducts in other areas will likely have the same effect as turning up the heat. It will also create cold working conditions in other areas of the plant, thus potentially creating a morale issue.

6. *Change the paint shift schedule to avoid the early morning.* Changing the paint areas shift schedule may reduce or eliminate the effect of the cold temperatures on the situation. This would create a host of scheduling issues for the other areas of the plant regarding the flow of the product. This could result in late delivery to customers.

7. *Require material handlers to take the long way around when receiving paint.* If material handlers are not opening the door, cold air from outside will not affect the paint flowing through the paint feed lines. This will also make a more pleasant working environment for the paint area associates.

Using the countermeasure evaluation table in Table 8.1, evaluate each of the seven remaining countermeasures using the information provided in this chapter.

Table 8.1   Countermeasure Evaluation Table

Countermeasure	Effectiveness	Feasibility	Impact
1. Relocate overhead door and block off opening.			
2. Direct space heaters at paint feed lines.			
3. Wrap feed lines in heat trace tape.			
4. Turn up thermostat in paint area.			
5. Close heat ducts in other areas.			
6. Change paint shift schedule.			
7. Material handlers take long route.			

Selecting Countermeasure(s)

After you complete the evaluation process, it is time to select the most suitable countermeasure(s) for your problem. There is rarely a single magic bullet that will solve your problem and allow you to achieve your targets. To select the correct combination of countermeasures, you must take several things into consideration:

■ Will the countermeasure(s) achieve or help achieve the target?
■ If the evaluation process revealed a countermeasure that will address the root cause of your problem and prevent recurrence, will it be necessary to implement a short-term countermeasure?
■ Does the countermeasure(s) truly address the root cause of the problem?
■ Will the countermeasure(s) prevent recurrence of the problem?

Table 8.2 reflects how the original group evaluated the seven countermeasures. The following explains why the group evaluated each countermeasure the way it did:

The group felt that the heat trace would be highly effective in maintaining the temperature of the paint in the paint feed lines. According to the maintenance supervisor, the cost of materials and the man-hours required to install the heat trace were minimal, making it highly feasible. Applying heat trace to the paint feed lines would not adversely affect anyone, thus resulting in a High rating for impact.

Asking the material handlers to take the long way around when unloading and receiving paint would be highly effective in keeping the overhead door closed, preventing the cold air from affecting the paint, especially if the overhead door were locked. Because there is no cost or time required to implement the countermeasure, it was perceived as highly

Table 8.2 Completed Countermeasure Evaluation Table

Countermeasure	Effectiveness	Feasibility	Impact
1. Relocate overhead door and block off opening.	H	M	M
2. Direct space heaters at paint feed lines.	H	H	L
3. Wrap feed lines in heat trace tape.	H	H	H
4. Turn up thermostat in paint area.	L	L	L
5. Close heat ducts in other areas.	L	L	L
6. Change paint shift schedule.	H	M	L
7. Material handlers take long route.	H	H	M

Note: H = High, M = Medium, L = Low.

feasible. Asking the material handlers to take the long way around would require that they be outside in the cold for a longer period of time; however, keeping the overhead door closed would reduce paint department rework and keep the paint area associates warmer. Due to the double-edged nature of this countermeasure, the group rated the impact Medium.

Removing the overhead door and blocking off the opening would be highly effective and prevent any cold air from entering the paint area. There is some cost associated with blocking off the wall, but the cost would be offset by the reduction in rework on parts with paint spits. The drawback to this countermeasure is that the work would have to wait until the weather warms up, thus making this moderately feasible. Blocking off the wall would provide a more comfortable working area for the painters but cause the material handlers to be outside longer, thereby making the impact Medium.

The space heater would likely be highly effective in keeping the paint from getting too cold, resulting in paint spits. The feasibility is High because the maintenance department currently has space heaters. The impact was thought to be Low because there is no way to control how much heat is applied to the paint feed lines, which may result in other problems.

The group felt that changing the paint department schedule to avoid the cold hours of the morning would be highly effective in reducing the problem of paint spits. The group did not feel this countermeasure would present any problems for quality, cost, or safety but they were not sure that management would approve the idea so they rated the feasibility as Medium. When it came to impact, the group felt it was Low because it would create more problems in scheduling not only for the company, but also for some of the associates who would need to try to find transportation and childcare.

Turning up the heat and blocking off heat ducts to force additional warm air into the paint area received the same ratings for effectiveness, feasibility, and impact. Both countermeasures would increase the warmth in the paint area but only as long as the overhead door remained closed, resulting in the group rating the effectiveness as Low. Feasibility was also rated Low due to the increased cost of utilities when the heat is turned up in the paint area.

Notice how the group placed a qualifier of placing a lock on the overhead door during the discussion regarding the countermeasure that would require the material handlers to take the long route. The group realized that requiring the material handlers to take the long route would not necessarily guarantee that the countermeasure would be followed. Keep in mind that as your group members evaluate countermeasures, they can make small refinements just as this group did.

Based on the results of the countermeasure evaluation table, there are three countermeasures that stand out; numbers 1, 3, and 7. To decide on which countermeasure(s) should be implemented, the group needs to ask several questions. These questions will guide the group in selecting the countermeasures that will provide the best possible combination of results for the situation.

- Are there countermeasures that will address the root cause of the problem?
  - Removing the overhead door and filling in the opening with brick will provide a barrier between the cold air and the exposed paint fill lines.
  - Locking the overhead door and requiring the material handlers to take the longer route will eliminate the overhead door from being left open while unloading drums of paint.
- Do any of the countermeasures prevent recurrence of the problem?
  - Removing the overhead door and filling in the opening with brick will prevent the cold air from entering the paint area.
  - Locking the overhead door and requiring the material handlers to take the longer route will prevent the cold air from entering the paint area.
- Will it be necessary to implement a short-term countermeasure?
  - Removing the overhead door and blocking off the opening: Yes, it will not be possible to block off the opening until warmer months without affecting production.
  - Requiring material handlers to take the long route: No, as long as the overhead door is locked and cannot be opened by material handlers, the cold air will be kept out.
- Are there short-term countermeasures that will act as a Band-Aid for those countermeasures that cannot be implemented immediately?
  - Applying heat trace to the paint feed lines will ensure that the paint does not get too cold, providing a stop-gap measure until the overhead door can be removed and the opening blocked.
- Will any of the countermeasures allow you to achieve the target?
  - Installing the heat trace will get you through the cold months until the overhead door can be relocated and the opening blocked.
  - Locking the overhead door and requiring the material handlers to take the long route when receiving drums of paint. Will eliminate the downtime due to cold paint.

The key to selecting the most suitable countermeasures lies in your experience and ability to get the facts about each countermeasure. As a problem solver, you must resist the urge to circumvent the problem-solving process.

I facilitated a problem-solving event in which the top-ranking person in the event had his own sense of the problem and what should be done to solve the problem. When it came time to conduct the countermeasure evaluation process, he decided that he would select the countermeasures based on his perception of

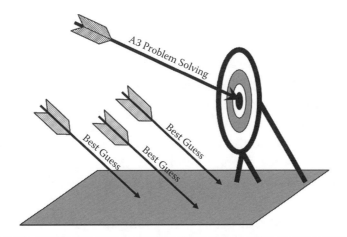

Figure 8.10 Process versus your best guess.

the problem and opinion of what would work. As it turned out, the results fell short of the expected target and he blamed the process.

The process will fail every time if you fail to follow the process. I call this approach the Best Guess method. Figure 8.10 shows how failing to follow the process can lead to failure. I can still hear David Verble saying, "Let the facts lead you." I cannot tell you how many times I have repeated those same words to others who would try to shortcut the process.

It always amazes me how many companies have the time and money to fix mistakes rather than do it right the first time. Problem solving is no different; you need to follow the process in order to have a real chance of reaching your targets and preventing recurrence.

The A3 Problem-Solving process not only provides you with a standardized method of determining what actions to take, but also provides a format for displaying your thought process. As a supervisor or manager, it is your responsibility to develop the people you supervise. The A3 is a great mentoring tool as it provides a format for guiding your people through the problem-solving process.

After talking with the material handlers, it was decided that locking the overhead door during the winter months and have the material handlers take the long route would be the best course of action. In addition, the group would request that management approve the purchase of several sets of coveralls to be used by material handlers when they needed to spend time outside moving materials.

In the Countermeasure(s) block of the A3, the author would list the short- and long-term countermeasures that were selected by the group. In addition, the author would write why the group is recommending the countermeasures that are listed. The recommendations are a critical part of the Countermeasure section of the A3. It allows the stakeholders to understand the thought process used to select the countermeasures listed in the A3.

Example

- *Short term:* No short-term countermeasure is required at this time as the long-term countermeasure can be implemented immediately and should not require any back-up measures.
- *Long term:* Lock the overhead door adjacent to the paint area and require material handlers to take the long route to the hazardous material room. Purchase several sets of coveralls to be used by material handlers.
- *Why recommended:* The long-term countermeasure will keep all associates from opening the overhead door during the winter months, and the coveralls will keep material handlers warm when they are outside. In addition, the cost of a padlock and several sets of coveralls will be much more cost effective than hiring a contractor to block off the opening of the overhead door. The long-term countermeasure will also keep the overhead door intact should it ever be needed to update or replace the paint booth.

During my years working at the training center at TMMK, we had thousands of people walking back and forth from the plant to the training center across the road. On those days when there was intermittent rain and people were not prepared, Team Members could get drenched on their walk to or from the training center.

During one such day, a group of people were talking about how to keep dry on rainy days when they had to come to the training center. Many ideas were expressed, such as building a covered walkway, installing an enclosed overhead walkway, and adding shuttle carts. Of course these ideas were very extravagant, thus making them highly unlikely to be approved. One person in the group came up with a different idea and submitted it through the suggestion program. The suggestion was to place umbrellas at the exit of the plant and at the entrance to the training center. This would allow people to take an umbrella from one of the locations any time it was raining.

This suggestion is a great example of effectively addressing the issue of staying dry through simplicity. Not all improvements cost large sums of money.

As an A3 Problem-Solving mentor, the objective is to develop the problem-solving skills of your associates so that it becomes a natural part of their thinking process. In the A3, there are three areas where you as a mentor can gain insight into the author's thinking process. The first is the *rationale* in the problem situation. The rationale will let you know if the author is selecting problems based on the needs of the company. The second area is in the *cause analysis* section where the author lists the potential causes of the problem and how the potential causes will be checked. In this section you will come to see how the author applies logic to guide the processes of trying to uncover the root cause of a problem. The third area is in the *countermeasure* section where the author explains why the countermeasures were recommended. The recommendations will tell you if the A3 author understands the balance between solving the problem and making informed decisions based on what is best for the organization as a whole. In the scenario herein, the author, with the help of the group and material handlers, selected a very low-cost action that could be implemented easily and solved the problem.

Chapter 9

Implementation

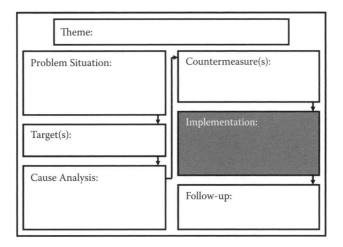

Figure 9.1   Implementation represents the DO phase and part of the CHECK phase of the A3 Problem-Solving process.

The 3-Cs for Success

Whether you are throwing a baseball or swinging a golf club, follow-through is a critical part of the process. It is not enough to identify the problem clearly; you must analyze to root cause and select countermeasures that will achieve your target. If you do not effectively implement the countermeasure(s), you not only will fail but you wasted valuable time and resources as well.

In the Implementation section of the A3 Problem-Solving process (see Figure 9.1), there are 3-Cs necessary to successfully implement your countermeasures:

- Create the plan.
- Communicate the plan.
- Carry out the plan.

Create the Plan

In the A3 Problem-Solving process, you must not take it for granted that once you have selected the best countermeasure(s) for the situation someone will take the lead in implementation. As the author of the A3, it is your responsibility to coordinate the implementation process to ensure success.

In creating the plan, there are three basic steps:

1. List required steps.
2. Sequence required steps.
3. Anticipate potential problems.

List Required Steps

The process of creating a plan to implement the selected countermeasure(s) begins by listing all the necessary actions required to implement each countermeasure. With the group, think through everything that must be done to get the countermeasure in place down to the smallest detail.

Example

James is the maintenance supervisor for one of several small manufacturing companies in Kentucky operated from a corporate office in Ohio. The Kentucky facility has been in business for the last 9 years. When the company first started, it hired experienced maintenance associates. Over the course of time, many of those experienced maintenance associates left the company, leaving the company short staffed in maintenance. The company decided 5 years ago to recruit associates from the shop floor and have them mentored by the remaining, experienced maintenance associates.

The problem is that, over the past 4 years, maintenance costs have increased due to contracting out much of the maintenance work at the facility. During the analysis phase, James determined that the cause of the increased maintenance cost was due to the lack of a set structure for mentoring the shop floor associates on the necessary maintenance skills.

The countermeasure James selected to address the problem is to develop and implement a new "pay-for-knowledge" maintenance training program in conjunction with the local community college. It is the first week in October and James hopes to get the new program approved by the end of the year so that money can be set aside for the program in the coming year's budget. The company shuts down the last full week of December. James's target is to have all maintenance associates trained on all skills within the next 2 years.

James came up with the following steps for implementing the new program:

1. Define required maintenance skills.
2. Prioritize the maintenance skills.
3. Write the policy for the program
4. Identify appropriate contact person at the community college.
5. Develop the training materials.
6. Conduct training on maintenance skills.
7. Communicate the program to maintenance associates and payroll staff.
8. Meet with community college representative to discuss the procedure.
9. Schedule classes.
10. Get policy approved.

Sequence Required Steps

Now that all the necessary steps for implementing the new program have been defined, they must be sequenced in order of implementation. To do this, James must think about what needs to be accomplished. He must also specify who will do what, when, where, and sometimes how the action will be implemented.

James began sequencing the necessary steps:

■ First, he had to find a way to define the required maintenance skills and then divided them into routine or challenging categories. He did this by looking through the maintenance department files at all the maintenance work orders for the past year.
■ Then he looked to see how each work order was handled (i.e., in house or by a contractor). James determined that defining the required maintenance skills needed to happen by October 5, 2009, before he met with the community college on October 7, 2009
■ James also wanted Heidi, the corporate HR director, and Janet, the plant training specialist, to be involved in the meeting. He booked Meeting Room 2 for the meeting because it has a videoconferencing system. This would make it possible for Heidi to attend the meeting without leaving the corporate office. The meeting was scheduled for October 13, 2009.
■ James asked Janet to have a draft of the pay-for-knowledge program ready by October 9, 2009, for Paul's review. Heidi would then walk the policy through the approval process no later than November 11, 2009.
■ James asked Bob at the community college to have the curriculum developed and ready for review and approval by March 8, 2010.
■ James and Janet would announce the new maintenance pay-for-knowledge program to all maintenance associates in the training room by February 2, 2010. This would enable them to find out if maintenance had any scheduling issues.

- Using Outlook, Janet would make sure that rooms were available for the training dates no later than February 16, 2010.
- Bob will teach the courses using a combination of classroom and shop-floor simulation. The first round will run from April through June 2010, the second round will run from September through November 2010 and the third round will run from February through April 2011.

To make it easy to read and understand, the list of sequenced actions should be placed in a simple table like Table 9.1.

Table 9.1 Implementation Table

What	Who	Where	How	Projected Completion	Actual Completion
1. Define required skills and prioritize	James	Maintenance dept.	Review work orders	10/5/09	
2. Identify contact and schedule meeting with community college	James	Plant	Telephone	10/7/09	
3. Meeting with community college	Heidi James Janet	Meeting Room 2	Face-to-face videoconference	11/11/09	
4. Draft program policy	Janet	Training dept.	N/A	10/9/09	
5. Get policy approved	Heidi	Corporate office	Face-to-face with President	11/11/09	
6. Develop curriculum	Bob	Community college	N/A	3/8/10	
7. Communicate program to Maintenance and Payroll	James Janet	Meeting Room 2	N/A	2/10/10	
8. Set class schedule	Janet	Plant	Outlook	2/16/10	
9. Train maintenance workers	Bob	Scheduled room	Classroom and simulation	Rd1 Apr–Jun Rd2 Sept–Nov Rd3 Feb–Apr	

When creating the plan, it is important to offset the schedule dates for implementing multiple countermeasures. James wants to be able to evaluate the effect that each countermeasure has on the problem. If he implements all the countermeasures at the same time, it will be impossible to know which countermeasures were effective.

Now James must ask himself "What if?" to anticipate possible problems with his plan. For example, what if the materials do not arrive on time or some materials are back-ordered? This process will allow him to make contingency plans should something go wrong.

Communicate the Plan

An important distinction to make at this point is that the Problem-Solving A3 format is a proposed plan of action. It must be communicated to your supervisor and all decision makers affected by the plan in order to gain the support needed for successful implementation. This process is referred to as "Nemawashi".

Nemawashi is a term I learned at Toyota; it means "to lay the ground work for major changes." This is accomplished by individually consulting those decision makers who are affected by the change, and in some cases, decision makers who are not affected. The purpose is to get their feedback and opinions on how you propose to address the situation. Based on the feedback from these individual discussions with the stakeholders, the author may need to modify the plan to address concerns to gain the needed support and approval to implement the plan.

Show your plan to anyone else who will be involved in or affected by the implementation of your countermeasure(s), including associates on the shop floor or in the office. The A3 should be posted in a conspicuous place in the work area so that everyone understands what the change will look like once all actions have been taken.

Carry out the Plan

Effective implementation cannot be assumed. It must be planned, and the plan must be followed to ensure effective countermeasure implementation. As the owner of the A3, it is your responsibility to ensure that all people in charge of action items follow the plan.

In my experience, it is not uncommon for individuals or organizations to implement an action item, policy, procedure, or Lean concept and then walk away with the feeling that they accomplished what they set out to do, only to find at a later date that the action was not carried out completely. For that reason, it is essential that you monitor the progress of the action plan to ensure timely implementation.

There are three things that the author of an A3 must do once the plan is created and approved:

1. Assessment
2. Assignment
3. Accountability

Assessment

The author must review the plan at regular intervals to make sure that each action item is being completed on time. The interval depends on the project or the status of the project. The tighter the deadline, the more frequent the checks should be conducted. The reviews should include "report-outs" from those responsible for implementing each item on the implementation plan.

Assignment

The implementation plan defines the assignments for all involved. During the assessment, you may identify additional steps that need to be completed or find that steps that were to be completed were not. It then becomes necessary to make new assignments or reassign certain steps.

Accountability

It is not enough for the people responsible to say that they ran out of time or other things came up. They need to set the excuses aside and tell you how they plan to get back on track so that the plan does not fall short. Posting the plan in a visible place frequented by management creates a little more urgency on the part of others, especially if there is some sort of regular performance evaluation process. This is why the Nemawashi process is so important. Gaining buy-in at the beginning makes it easier for the author to keep everyone focused.

During the process of creating the plan, we discussed anticipating problems so that if things did not go as planned, you would be prepared to take contingency actions. When things go awry, you must be prepared to modify the plan to keep the problem-solving effort on track. This also means that you will have

Table 9.2   Implementation Table with Actual Dates

What	Who	Where	How	Projected Completion	Actual Completion
1. Define required skills and prioritize	James	Maintenance dept.	Review work orders	10/5/09	10/3/09
2. Identify contact and schedule meeting with community college	James	Plant	Telephone	10/7/09	10/7/09
3. Meeting with community college	Heidi James Janet	Meeting Room 2	Face-to-face videoconference	10/13/09	10/10/09
4. Draft program policy	Janet	Training dept.	N/A	10/9/09	10/9/09
5. Get policy approved	Heidi	Corporate office	Face-to-face with President	11/11/09	10/12/09
6. Develop curriculum	Bob	Community college	N/A	3/8/10	(3/15/10)
7. Communicate program to Maintenance and Payroll	James Janet	Meeting Room 2	N/A	2/10/10	2/10/10
8. Set class schedule	Janet	Plant	Outlook	2/16/10	2/9/10
9. Train maintenance workers	Bob	Scheduled room	Classroom and simulation	Rd1 Apr–Jun Rd2 Sept–Nov Rd3 Feb–Apr	

to keep all individuals, groups, and departments that are participating in the implementation apprised of any changes to the plan.

Table 9.2 shows the actual date an action was taken versus the projected date of completion. Anytime a projected date is missed, you should be prepared to explain why the date was missed and what you intend to do to get back on track.

Chapter 10

Follow-up

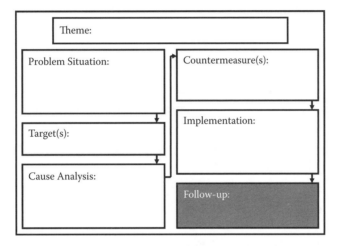

Figure 10.1   The purpose of Follow-up is to track and visually display the results of the implementation plan.

Evaluating the Results

The ultimate aim of A3 Problem Solving is to achieve the stated target(s). Once the implementation is carried out, you must evaluate the results of the countermeasure(s) to determine the degree of success or failure that each countermeasure had on the problem by measuring progress toward the target(s).

The purpose of Follow-up (see Figure 10.1) is to track and visually display the results of the implementation plan. Follow-up consists of three components:

1. How to check
2. When to check
3. Recommended actions

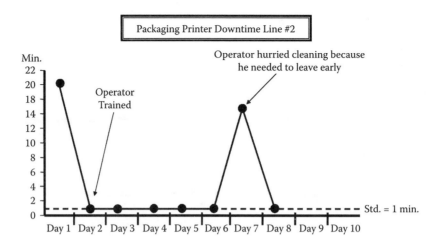

Figure 10.2 Follow-up line graph.

How to Check

You will need to decide how you are going to measure the effectiveness of the countermeasure, and then project over time the effect that each countermeasure will have on the problem situation. The most logical form of measurement would be in the same form as your problem-solving target. Consider the example given earlier in the text:

> *Target:* Reduce printer downtime due to clogged spray nozzle on line #2 to 1 minute or less.

To effectively determine the success or failure of your actions, you should be tracking the number of minutes line #2 is down due to a clogged printer nozzle. You will also need to determine at what intervals you will track the information. Figure 10.2 is a simple line graph showing actual results with explanations.

When to Check

Deciding when to check should be based on the milestones identified during implementation and the way the standard is expressed in the problem situation. If the standard for downtime is expressed daily, then hours or minutes would be appropriate. If your problem is scrap and it is tracked monthly, you may want to track it weekly or even daily to see if your actions are moving you in the right direction instead of waiting until the end of the month when it is too late. It is also helpful to identify the milestones on your Follow-up graph. This will allow you and others to see what effect each countermeasure had on the problem.

Table 10.1 provides a check sheet that associates can complete to track downtime as they perform their tasks.

Table 10.1   Scrap Check Sheet

Packaging Downtime				
	Line #1	*Line #2*	*Line #3*	*Total*
6:00 to 7:55				
1st Break				
8:05 to 10:00				
Lunch				
10:30 to 12:25				
2nd Break				
12:35 to 2:30				

Table 10.2   IF-THEN Table

Recommended Actions	
IF	*THEN*
Countermeasure implementation plan failed	Begin A3 Process to identify root cause of the failed plan
Countermeasure(s) partially achieved the target	Identify other root causes and find new countermeasure(s)
Target was achieved but problem could recur	Look for other countermeasures that will prevent recurrence
Target achieved and problem will not recur	Standardize countermeasure(s) and share findings
Problem solved	Pick up new problem

Recommended Actions

After gathering data on how well the countermeasure(s) worked, you need to decide what additional actions need to be taken in relation to the problem. This means evaluating the results of your countermeasure(s).

There are four basic things you need to know in order to make the necessary Follow-up decisions:

1. Did the implementation plan fail?
2. Was the target partially achieved?
3. Was the target completely achieved?
4. Will the countermeasure prevent the problem from recurring?

Table 10.2 is an IF-THEN table to help you determine your next steps.

Chapter 11

Putting It All Together

Dave's Fabrication Case Study: Part A

Monday, May 10, 2010. Dave's Fabrication is a family-owned job shop in Dallas, Texas, that produces small parts for manufacturers of medium- and heavy-duty equipment. The company is in the process of transforming from a traditional approach of manufacturing to one that focuses on reducing waste in the manufacturing process.

This transformation is due, in large part, to customers demanding their product in shorter periods of time. Most of the workers on the shop floor have been with the company for at least 15 years. The expectation that they are supposed to bring problems to leadership's attention is a foreign concept for most of them, and they are having a difficult time making the transition from an environment where passing the buck was okay. Most of the workers are accustomed to doing things a certain way and rely on their leaders to tell them what to do when something goes wrong. When asked why something is done a certain way, the reply is often, "That's the way we've always done it."

It is Monday morning and Paul, the Production Control Supervisor, met with Holly, the Production Supervisor, to discuss results of the job reports for the past 2 weeks. These reports compare the quoted hours with actual hours per job. The reports for Holly's department show, for the second week in a row, that the backrest process exceeds the budgeted hours for the job. The report shows that it took 24.75 hours to weld 90 backrests, which is 1.5 times longer than the original quote of 16.5 hours. Robby, the backrest welder, has to make 22 welds that are 2 inches long to complete one backrest. The quote was based on 0.25 minutes per inch.

You know the following information:

- Robby, a long-time employee, has only been working in the welding department for 2 weeks. Kelly, the welding Team Leader, told you that she had the best welder in the department teach Robby how to do the job properly.

■ On Monday, May 24, 2010, the customer's batch size will go from 90 back-rests to be delivered once per week on Fridays to 90 per batch to be delivered on Tuesdays and Thursdays, for a total of 180 per week.

■ Flat steel that is 2 inches wide by 0.25 inch thick is sheared to length in the Machining department. It then moves to a computer numerical controlled (CNC) machining center where the edges are beveled. The pieces are then delivered to the backrest department where they are welded to form backrests.

■ All shop-floor departments work Monday through Friday from 7:00 a.m. to 3:30 p.m. The office works Monday through Friday from 9:00 a.m. to 5:30 p.m. Both shop and office employees work 8 hours per day with two 10-minutes breaks and one 30-minute break. Based on the current situation to meet the new demand rate, the backrest area will have to work from 7:00 a.m. to 4:30 p.m. Monday through Friday.

■ You are concerned because, in addition to the increased cost per unit, it will take 49.5 hours at the current rate of production to complete 180 backrests. You only have 2 weeks to meet the customer's new demand rate.

■ The Purchasing department has informed you that it has scheduled a meeting with a representative of the steel company with which you started doing business in February of this year. That company was selected because it offered a substantially cheaper price than the company you did business with previously. Purchasing conducts a 90-day review with all new suppliers and has asked that you attend to provide feedback on the quality of the materials.

■ Holly asked Kelly to use her newly learned A3 Problem-Solving skills to define the problem, determine the root cause(s), and develop and implement countermeasures to address the problem. To get Kelly started on the right foot, she reviewed the results of the job reports for the past 2 weeks.

> **Instructions: Using the information in Part A of the case study and the blank A3 entitled "Dave's Fabrication," complete the Background, Standard, Current Situation, Discrepancy, and the Rationale of Appendix O.**

Dave's Fabrication Case Study: Part B

Kelly decided to talk with Robby to get more information about the problem so she could break it down into specific characteristics. Robby explained that he had been trained to make a Tee Weld to join the slats to the frame of the backrests. He further explained that the day after he was put on his own, the slats did not fit like the ones he used during his training. Robby showed Kelly how the slats seemed to be angled on both ends, leaving a gap between the slats and the outer frame of the backrest.

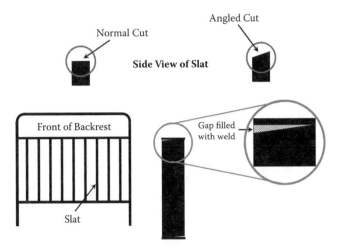

Figure 11.1   Forklift backrest and slats.

Robby explained that because of the gap, it took longer to join the slats to the backrest frame. Kelly asked why he had not said anything to her about the change in material. He stated that he did not want to complain the first day on his own and did not mind the additional challenge. Figure 11.1 is a drawing of a forklift backrest and side view of the slats.

Now that Kelly had a better understanding of the problem, she decided to track back through the processes to find the Point Of Cause (POC) for the angled cuts. Her first stop was the Machining department, where she explained to Marty (the department Team Leader) the problem she was having. Together they checked the slats at the CNC machining center where all the edges on the slats are beveled. After inspecting the machined slats and slats entering the department to be machined, they found that both sets of material were angled.

They then moved on to the shearing process where flat stock is sheared to length in order to make the slats for the backrest. They inspected the raw flat stock at the shear to see if the material was coming from the supplier with angled ends. Using squares, they inspected all the stock and found it to be square. They then checked the pieces coming off the shear and found that they were angled at both ends. Together, Kelly and Marty were able to confirm that the shear process was the POC for the problem.

After thinking through everything, Kelly realized that in order to meet the customer's new schedule, she would have to have the backrest welding time back down to 0.25 minutes per inch by Monday, May 17.

Instructions: Using the information in Part A and Part B of the case study, complete the Extent section of the Problem Situation and refine the Problem Statement. In addition, complete the Target statement and Theme sections of Appendix O.

Dave's Fabrication Case Study: Part C

Kelly asked Marty what could cause the shear to produce angled cuts. Marty explained that if the shear operator did not hold the flat stock flush against the shear's squaring arm, then parts would come off with an angled end. Marty also explained that it would be unlikely that this could be the cause of the problem because the degree of the angle on the slats is the same on every slat. If the problem was due to the operator not holding the part flush on the squaring arm, the angles would most likely vary in degree.

Another potential cause could be that the shear's blade could be out of alignment. If the blade is out of alignment, it would cause angled cuts on the ends of the backrest slats.

The only other potential cause Marty could think of was that the squaring arm might have been misaligned. A misaligned squaring arm would also create an angled cut on the ends of the backrest slats that would be consistent from one part to the next.

Kelly and Marty decided to observe Bill, the shear operator, from a discreet distance as he cut slats for the next day's production. They noticed that he pushed the slats flush against the squaring arm every time he cut slats. They waited until Bill went on his afternoon break to check the alignment on the shear blade and squaring arm. First they took three measurements (left, right, and center) from the front of the shear to the front edge of the shear blade to see if it was squarely aligned. They compared all three measurements to the readings taken when the shear underwent Preventive Maintenance (PM) 5 weeks earlier. All three measurements were exactly the same and matched the PM results.

Next they looked at the squaring arm; it appeared to be square. They placed a recently calibrated square flush against the front edge of the shear blade and then slid it up against the squaring arm. There was a slight gap between the square and the squaring arm. This gap confirms that the squaring arm is misaligned. They checked the results of the last PM and found that the squaring arm was judged to be square. Because the slats had only been out of square for the past 2 weeks, they decided to check the shear for signs of damage. They noticed that there was some paint at the very end of the squaring arm, which is consistent with the orange paint found on forklifts driven by Material Handling Team Members.

When the shear operator returned from break, Marty and Kelly asked him a series of questions to try to find the root cause of the problem.

Marty: Bill, do you know how the orange paint got on the end of the squaring arm?

Bill: About 2 weeks ago, a material handler was having difficulty backing out of my process area and the counterweight brushed up against the squaring arm. The area is pretty congested, so there is not a lot of room to maneuver a forklift.

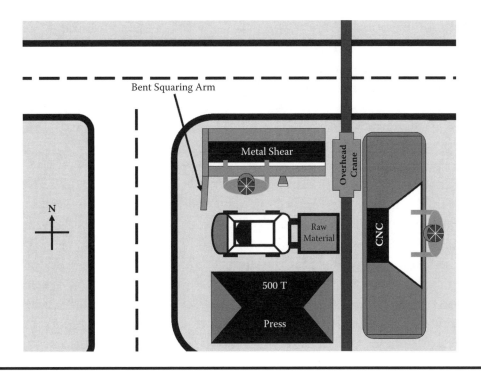

Figure 11.2 Shear area layout.

Kelly: Why was the material handler in the process area with a forklift?

Bill: He was delivering materials that needed to be sheared.

Marty: Bill, are you aware that material handlers are supposed to use the overhead crane (see Figure 11.2 shear area layout) to position materials in process areas due to the potential for personal injury and equipment damage?

Bill: Yes, but in the 10 years that I've worked here, they've always delivered parts and materials using a forklift.

Marty: Why didn't you say anything about the accident?

Bill: I looked it over and, other than the orange paint, it looked like it was okay. No harm, no foul.

Kelly explained the problem that the backrest welding department has been experiencing for the past 2 weeks and that it was tracked back to the shearing process. She further explained how something as seemingly harmless as the forklift brushing up against the squaring arm could cause a more serious problem down the line.

Marty: Bill, it is important that you tell me anytime something out of the ordinary happens that could affect production.

Bill: Okay.

Kelly: Marty, is there any other way to cut backrest slats other than with the shear?

Marty: Before we got the shear 5 years ago, we cut all the slats on a band saw. It was a little slower but it made really good parts.

Kelly: Would it be possible to use the band saw to cut slats until the squaring arm can be repaired?

Marty: Not a problem. Tomorrow I'll have Bill start cutting any critical parts on the band saw until the shear is repaired. By tomorrow, I will have a purchase requisition approved for a contractor to come in and repair the squaring arm on the shear. If the contractor is not too busy, I should be able to have it completed by the end of the month.

Tuesday, May 11. First thing the next morning, Kelly stopped by the dock area to talk to Diane, the Material Handling Team Leader. She hoped by talking with Diane that she could get to the root cause of the problem. She explained the problem and what she had found thus far.

Diane called Mark, the Material Handling Team Member who is responsible for delivering to the shear, and asked him to meet with her and Kelly at the docks. During the conversation with Mark, they found out that he did brush up against the squaring arm on the shear but did not report it because the operator had said there was no damage. After establishing that Mark was the forklift driver responsible for the misaligned squaring arm, they had a few more questions for him.

Diane: Mark, why do you drive the forklift into the process area instead of using the overhead crane as you were instructed?

Mark: It just doesn't make any sense to me to get off the forklift and hook the parts bin to the crane when the forklift can fit into the area.

Diane: Do you know why you are not supposed to drive forklifts into process areas?

Mark: No, I was just told to use the crane, but it was never really explained to me.

Diane: I'll explain it to you when we are done here. During tomorrow's team meeting at the start of the shift, I will review the Job Instruction with all material handlers. I'll be paying particular attention to the safety key points so that this does not happen again.

> **Instructions: Using the information in Parts A through C of the case study, complete the Cause Analysis section of Appendix O.**

Dave's Fabrication Case Study: Part D

Feeling confident that the root cause had been identified, Kelly began thinking of everything that needed to be completed to ensure that all necessary actions would be taken to address this problem and keep it from recurring.

Today, Diane is going to explain to Mark the safety concerns of operating a forklift in a process area. Tomorrow, she will cover the same information at the team meeting at the beginning of the shift. Instructing all material handlers on the safety key points will address the root cause of the problem and should pre-vent any further issues.

Today, Bill is supposed to start cutting all backrest slats on the band saw. Also today, Marty will have a purchase requisition approved for a contractor to fix the squaring arm by the end of the month. This action will ensure that slats have square ends and eliminate the extra welding time required to fill the gaps.

Kelly talked to Marty today and they both agreed that erecting a guardrail along the west side of the shear area would be a good idea. Kelly agreed to submit the maintenance work order today. Maintenance promised that Willie and Eve would have it erected no later than Thursday of this week. Erecting a guard-rail will prevent forklifts from entering the shear process area.

> **Instructions: Using the information in Part D of the case study, complete the Countermeasures and Implementation sections of Appendix O.**

Dave's Fabrication Case Study: Part E

After completing the Implementation section of the A3, Kelly met with Holly to review the results with her and get her buy-in and approval. Holly was very complimentary of Kelly's hard work and thanked her for doing such a good job of letting the facts influence her decisions. She told Kelly that she would need to review her results with Paul because he would need to approve the Purchase Order (P/O) for the contractor.

Paul was very impressed by Kelly's analysis of the problem and commented on how easy the A3 made it to understand how she got from long welding times on backrests, to the material handler not understanding why entering process areas is prohibited. Paul asked Kelly what she was going to do next. Kelly replied, "I need to implement my action items and make sure that I track the progress of my implementation plan. I also need to track the average daily time required to weld backrests to ensure that the problem is solved." The results were as follows:

- Implementation plan results:
  - Instruct Mark on safety key points – Actual date = May 11
  - Instruct all Material Handlers during team meeting – Actual date = May 12
  - Begin to cut slats on band saw – Actual date = May 11
  - Approve P/O for contractor to realign squaring arm – Actual date = May 11
  - Contractor realigns squaring arm – Actual date = May 25
  - Submit guardrail Work Order (W/O) – Actual date = May 11
  - Erect guardrail – Actual date = May 17 – Materials were not available and needed to be ordered.

■ Welding time results:
  – May 11: Average backrest weld time = 16.5 minutes
  – May 12: Average backrest weld time = 11 minutes
  – May 13: Average backrest weld time = 10 minutes
  – May 14: Average backrest weld time = 10 minutes
  – May 17: Average backrest weld time = 10 minutes
  – May 18: Average backrest weld time = 10 minutes

Kelly is also going to recommend the installation of guardrails in all areas where there is a potential for forklifts to enter process areas.

> **Instructions: Using the information in Part E of the case study, complete the Implementation and Follow-up sections of Appendix O. After completing Appendix O, you may compare your A3 to the examples in Appendix P, Dave's Fabrication Answer.**

Chapter 12

Improve Your A3's Effectiveness

Sequencing the Flow of an A3

Following the A3 Problem-Solving process is not enough if you do not document the information in such away that makes it easy to follow and easy to understand. This chapter explains some simple and effective ways to help you convey your problem-solving story. It also explains the importance of how the information in your A3 flows, how certain visuals can enhance the reader's comprehension, and provides an opportunity to practice creating some visuals.

When it comes to flow, I have heard others say that the left side of the A3 explains the problem and why it happened, and the right side explains what you intend to do about the problem. This is not necessarily the case. Instead of talking about what goes on each side of the A3, I would like to explain how the information should flow.

Many places I visit do not have standard forms for information that is frequently communicated. This lack of standardization makes it difficult for those who need to make sense of the information. They spend additional time scanning documents to find the needed information. To avoid this type of confusion, I provide a standardized A3 to my class participants to use for their class project. The standardized A3 provides a flow that is easy to follow and develops a common format for everyone in the organization. Figure 12.1 is an example of an A3 with poor flow.

The best way to think about laying out an A3 is to imagine that the 11- by 17-inch sheet of paper is actually two 8.5- by 11-inch sheets of paper sitting side by side. You must fill up one piece of paper before you move to the next. If you think of it like that and follow the process block by block, then you should not have any problems with flow or problems with your readers getting lost.

Figure 12.2 shows a Problem-Solving A3 format that does not look like most of the A3s in this text. However, it does follow the A3 Problem-Solving process.

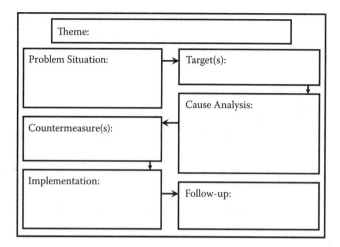

Figure 12.1   Example A3 with poor flow.

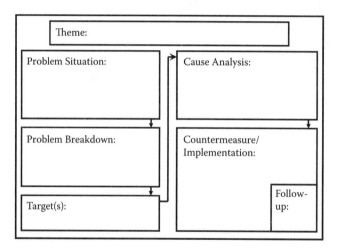

Figure 12.2   Proper A3 flow.

As you can see in this example, the Cause Analysis or information regarding why the problem happened is on the right side of the A3, not the left. The most important thing to remember about laying out an A3 is the flow of information, not on what side the information falls. If you worry about where the information is supposed fit, you may sacrifice necessary information.

Common A3 Visuals

For the past 17 years, I have incorporated a "learning styles" assessment into my Job Instruction class. It illustrates that as individuals we all have different ways in which we learn. Some people learn more effectively through their auditory senses. Others prefer relying on their visual senses to other methods for learning, while yet another group would much rather learn through a hands-on approach.

The assessment allows the class participants to recognize that, although each person has a preferred method of learning, each person uses all three senses (auditory, sight, and touch) to learn, depending on the situation. Through the years, I have tracked the results of the assessment and found that more than 60% of the class participants indicated sight as their primary means of learning.

We have all heard the saying that "a picture is worth a thousand words." Using visuals allows you to convey the same information in less space and creates a more visually appealing A3. If an A3 is completely covered in text, it becomes very overwhelming to the reader. By using visuals, you are able to free up space on the A3, creating more white space and making it easier to read.

If you are anything like me, you prefer to look at the pictures instead of reading paragraph after paragraph of instructions when assembling a new toy or piece of furniture. So, it stands to reason that by incorporating visuals into an A3, you will improve the reader's ability to understand the information you are trying to convey.

The most common visuals used in A3 Problem Solving include the following:

- Line graphs
- Data tables
- Pareto charts
- Pie charts
- Pictograms
- Sketches and drawings

This section explains each type of visual and also provides examples. Before delving into each of the visuals, let's discuss some basic equations that will help in the creation of the visuals.

Basic Equations

Because the production requirement can change from day-to-day, week-to-week, or month-to-month, it is necessary to convert numbers of occurrences or defects to a defect rate or percent. Converting the number will allow you to make direct comparisons from one time period to another.

Convert a number to a defect rate. Divide the number of occurrences or defects by the daily, weekly, or monthly production requirement to establish the defect rate.

Weekly production requirement = 1,000 units
Number of defects for the week = 20
Standard = 0.015
Conversion to a defect rate: 20 defects divided by 1,000 units = 0.02

Note that the number of places you display to the right of the decimal point depends on your standard. If the standard displays three places to the right of the decimal point, then you need to maintain the same number of places to the right of the decimal point when calculating defect rates. For example,

Weekly production requirement = 11,290 units

Number of defects for the week = 10

Standard = 0.0003

Conversion to a defect rate: 10 defects divided by 11,290 units = 0.0008857

Because the standard displays four places to the right of the decimal point, the answer would be 0.0009

If you left it as 0.0008 instead of rounding to 0.0009, you would be underestimating the current situation

Convert a number to a percent. Follow the same process as you did to convert a number to a defect rate. Then move the decimal point two places to the right.

Conversion to percent: 20 defects divided by 1,000 units = 0.02

By moving the decimal point two places to the right, you obtain 2%

During the Pareto exercise, you will need to convert defect rates to numbers; to do this, you must reverse the calculation

Multiply the defect rate times the daily, weekly, or monthly production requirement

Defect rate (0.0021) times weekly production requirement (10,000) = 21 defects for the week

You will need to use these calculations when completing the exercises that follow in the next section.

Line Graphs

Of the most commonly used visuals, the line graph is the most popular and used throughout the A3. It is often used in the Problem Situation to show the current situation as it relates to the standard or expectation. It also visually shows if the situation is getting worse, better, or staying the same—otherwise known as the trend. The line graph is also used in the Target section to show what the reader wants to accomplish by a specific date. It is also used in the Follow-up section to show the effects of the countermeasures over time.

Line graphs are used to show how a variable or a group of variables changes over time. The first axis (x-axis) is a horizontal line showing time. The increment of time depends on what you are measuring and how frequently you need to know the status of that information. The second axis is a vertical line that displays the scale (temperature, dollars, defect rates, etc.) and typically the unit that you will be using as the measure. This vertical line is also known as the y-axis. How you number your scale can dramatically affect how the reader perceives your information.

Figures 12.3, 12.4, and 12.5 provide examples of line graphs depicting the same information. The only difference is how the scale is numbered.

You need to create line graphs that will attract the attention of the stakeholders from whom you will ultimately need to gain approval. The key is not to

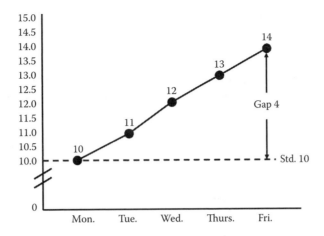

Figure 12.3   Exaggerated line graph.

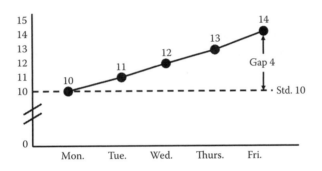

Figure 12.4   Normal line graph.

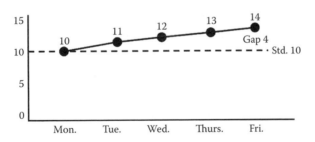

Figure 12.5   Understated line graph.

exaggerate or understate the problem. Because the idea is to show as much information visually as possible, be sure to show the value of each plot on the graph.

Data Tables

Data tables help you organize all your information in one place so that you can see patterns. The information is presented in a grid or matrix composed of columns and rows. The very first step is to create a title for the table that tells the reader what the data are about. For example, if you have class surveys on each of your instructors for a Quick Changeover class, the title of the data table could be "Quick Changeover Instructor Effectiveness."

Now that you have the title for your data table, you need to designate row (blocks running horizontally) and column (blocks running vertically) headings. Because you are trying to compare each instructor's effectiveness at delivering Quick Changeover classes, you will need to list all the instructors by name or instructor number in either the far left column or top row of the data table. For this example, the names will be placed in the far left column of the data table. The evaluation criteria found on the class surveys will be placed in the top row of the data table. Place the scores for each instructor in the corresponding rows and columns.

If you titled the data table and each of the rows and columns, then your data table should be easy to read. When using data tables in A3 Problem Solving, it is a good idea to incorporate the standard or expectation into the table. Once you have made your table, you can look for patterns in the data and draw some conclusions based on these patterns.

Data tables work well when there is a lot of information that you need to communicate in a small space. For that reason, data tables are commonly used in the Problem Situation block of an A3 to display multiple current situations as they relate to their respective standards. Highlighting where the current situation differs from the standard provides the reader with a much more comprehensive understanding of the problem situation. Table 12.1 is a completed data table for the Quick Changeover class surveys.

Pareto Charts

In addition to the following text describing the Pareto there is an animated PowerPoint presentation on the accompanying CD. The PowerPoint presentation shows how to construct a Pareto and how to use the Pareto to determine the extent of the problem.

A Pareto chart displays occurrences or defects from greatest to least, left to right, and looks like a bar chart with a line graph overlay. In A3 Problem Solving,

Table 12.1 Quick Changeover Instructor Effectiveness

Instructors	Criteria			
	Subject Knowledge	Well Organized	Communicate Content	Professional Attitude
James J.	4.75	3.25	5	4.9
Janet H.	4.5	3.75	5	4.9
Elaine L.	3.25	3.75	4.5	4.8
Jim E.	4.5	3.5	4.1	5
Standard	4.5	4	4.2	4.8

several Pareto charts can be used to break down the problem. If you only use one Pareto and focus on the largest category, you can easily be overwhelmed, making it difficult to succeed in your problem-solving effort. As you progressively narrow your focus through successive Pareto charts, it is critical to compare the data to the standard in order to know how far to narrow your focus.

A Pareto chart has a left-side vertical axis that measures the total number of occurrences. If you have five categories that you wish to display on a Pareto chart and the total of all five categories equals nineteen, then you must number the left side equal to that total. A Pareto chart has a horizontal axis at the bottom where the category of occurrences is displayed. It also has a right-side vertical axis that measures the accumulated percentage of all occurrences. The right-side axis is numbered from 0% to 100%.

Once you have the basic structure of the Pareto chart, you need to draw the bars that represent the occurrences you have identified. Label the top of each bar with its value.

Now that the bars are in place, you can draw the cumulative line. The cumulative line starts at 0% on the left side of the Pareto chart and works its way across the Pareto until it reaches 100% on the right side of the Pareto.

After constructing the Pareto, you need to figure out how many categories are needed to focus on in order to achieve the standard. To do this, you need to know the discrepancy described in number of defects. To convert a defect rate to a number of defects, multiply the defect rate by the requirement.

Example

Standard = 0.006
Current Situation = 0.014
Discrepancy = 0.008
A Standard (0.006) times the Production Requirement (1,000 units) = the Standard in number of defects (6)
A Current Situation (0.014) times the Production Requirement (1,000 units) = the Current Situation in number of defects (14)
The Discrepancy (0.008) times the Production Requirement (1,000 units) = the Discrepancy in number of defects (8)

This piece of information shows that you must focus on at least eight defects in the Pareto to achieve the standard. Starting at the left side and working your way to the right side of the Pareto, shade in the bars until you have at least eight defects highlighted. The shaded categories become your focus area.

After identifying the focus area, you will need to indicate the percentage the focus areas is of the total number of defects. This is done by drawing a line from the appropriate plot on the cumulative line and the right-side axis of the Pareto chart.

Once you have defined your focus area, you may need to break down the focus area further to identify the characteristics of the problem. This will require

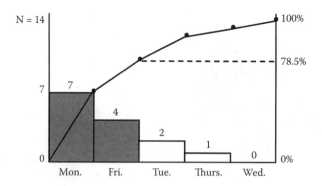

Figure 12.6   Pareto chart by days.

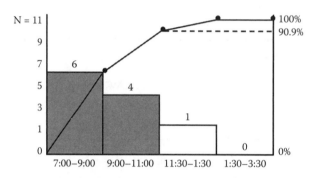

Figure 12.7   Pareto chart by hours.

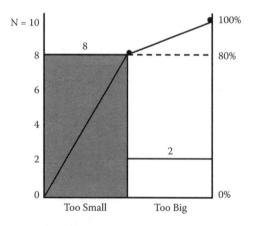

Figure 12.8   Pareto chart by size.

the creation of additional Pareto charts. Figures 12.6, 12.7, and 12.8 represent a sequence of Pareto charts that break down the problem to specific characteristics that can be analyzed to root cause.

Based on the information in the Pareto chart, you should begin asking why parts made on Monday and Friday between the hours of 7:00 a.m. and 11:00 a.m. are too small. The Pareto chart is a very valuable tool when it comes to breaking down the problem into something more manageable.

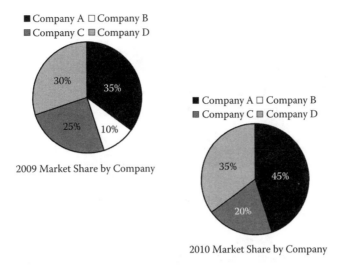

Figure 12.9 Market share pie charts.

Pie Charts

Pie charts are simple, round charts divided into sections that display the percentage of each variable as it relates to the whole pie. A pie chart is a good choice if all you want to do is show the current situation. Although the chart itself displays multiple variables, there is no direct means of comparing the information to the standard for each variable, thus making it difficult to determine if there is a problem without including additional visuals or text. Also, you cannot see the trend for a given situation unless you use multiple pie charts. If you want to show your company's market share in relation to your competitors' over the past 2 years, you would need to prepare two pie charts. Figure 12.9 shows the market share trend for several companies using two pie charts.

Pictograms

A pictogram is similar to a bar or column chart, but instead of using bars and columns, it uses pictures that are relevant to the data being communicated. The pictures are assigned a value and then stacked to the appropriate height depending on the data. The pictures create more interest and draw the attention of the reader more so than a simple bar or column chart. Like with any graph, a legend will help the reader understand the content, especially because partial pictures will represent fractions of the whole. Figure 12.10 is a pictogram of 1.5-ton forklift sales over a 4-year period.

Sketches and Drawings

There are times when, no matter how well you think you have described something, others will still have a difficult time making sense of it. Sketches and drawings make the situation come alive far better than words. With today's

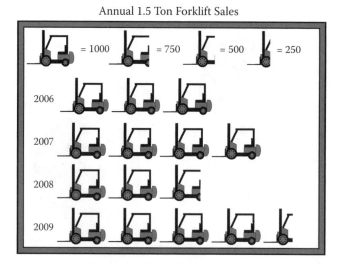

Figure 12.10   Pictogram of forklift sales.

technology, digital pictures can also be used to get across your point more effectively.

One example that comes to mind is a proposal that described the maintenance needs between two different types of stamped concrete bricks. A neighborhood association wanted to keep the unique look of bricked intersections that set it apart from other neighborhoods. The intersections were poured with concrete and then a rubber mat was pressed into the concrete to give it the look of brick. One stamp created a pillow-topped brick and the other created a flat-topped brick. The pillow-topped brick, because of its profile, created areas where snowplows could catch and damage the concrete. The flat-topped brick was flush with the surrounding asphalt, making it easier for the snowplow to glide over the top without catching. Not everyone easily understood the written description provided for the two types of stamps. Figure 12.11 is a simple drawing of the two types of brick. The drawing allowed the neighborhood association to see how the pillow-topped bricks could more easily be damaged by snowplows in the winter months.

All six visuals described in this text can be used in your A3s to improve your ability to tell the story. If you follow the A3 Problem-Solving process and develop actions that will address the root cause and prevent any recurrence of the problem, it is all for naught if you cannot capture the reader's attention. In addition, using visuals will allow you to create white space on your A3, and the white space will make the A3 more appealing to the eye.

Visuals Practice

After writing the original Problem-Solving materials at TMMK, our group created the next-level, problem-solving class called Problem-Solving QC (Quality Control) tools. Our Japanese instructor gave us a table with a list of defects and asked

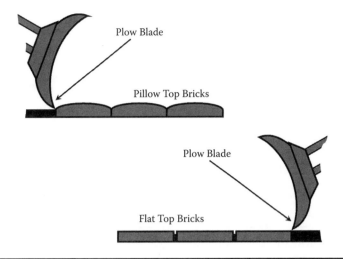

Figure 12.11 Drawing of two types of bricks.

us to create a Pareto chart of the defects. With a puzzled look on our faces, we began drawing. After 15 minutes, the instructor asked each of us to show the class our Pareto charts. Because we were all clueless as to what a Pareto was, we drew pie charts, tables, and bar charts. I am providing some examples of the tools so that you can refer back to them as you create your own. There are exercises for each type of visual except the sketches and drawings. After you complete the exercise, you will be able to find the model answer for each in the appendices at the end of the text.

Line Graph Exercise

You supervise an area that has Blow molding units, rotational molding units, and injection molding units. The weekly production for each area is 2,000 pieces. Chart the following information using the three blank line graphs in Figures 12.12, 12.13, and 12.14 and highlight the problem, if any:

Standard: The standard for all three areas is no more than 0.025 defects per week.

Current Situation:

■ Blow Molding: Week 1 = 20 defects, Week 2 = 30 defects, and Week 3 = 20 defects
■ Rotational Molding: Week 1 = 30 defects, Week 2 = 40 defects, and Week 3 = 40 defects
■ Injection Molding: Week 1 = 60 defects, Week 2 = 70 defects, and Week 3 = 90 defects

Discrepancy:
See Appendix J for answers to this exercise.

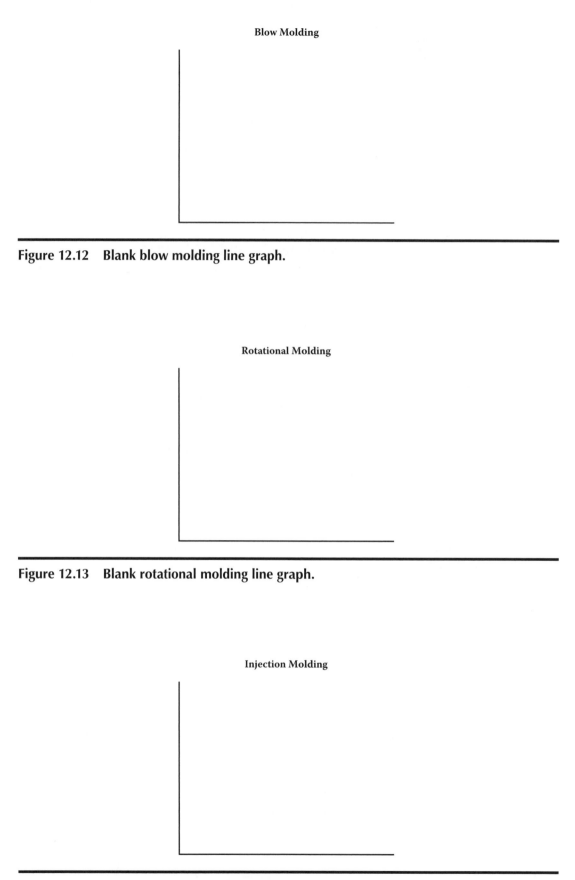

Blow Molding

Figure 12.12   Blank blow molding line graph.

Rotational Molding

Figure 12.13   Blank rotational molding line graph.

Injection Molding

Figure 12.14   Blank injection molding line graph.

Data Table Exercise

Using the blank space on this page, draw a data table that includes all the same information as in the Line Graph exercise.

- Standard: The standard for all three areas is no more than 0.025 defects per week.
- Current Situation:
  - Blow Molding: Week 1 = 20 defects, Week 2 = 30 defects, and Week 3 = 20 defects
  - Rotational Molding: Week 1 = 30 defects, Week 2 = 40 defects, and Week 3 = 40 defects
  - Injection Molding: Week 1 = 60 defects, Week 2 = 70 defects, and Week 3 = 90 defects

See Appendix K for answers to this exercise.

Pareto Chart Exercise

Use the following data to draw a series of three Pareto charts that break down the problem by machine (Figure 12.15), parts (Figure 12.16), and defects (Figure 12.17).

Machine Pareto Chart

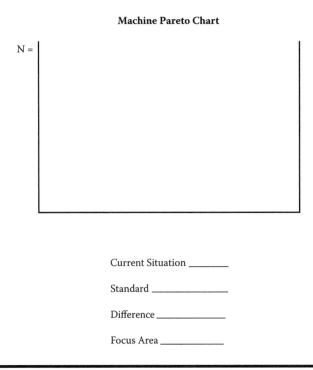

Current Situation _____

Standard _____

Difference _____

Focus Area _____

Figure 12.15   Blank machine Pareto.

Parts Pareto Chart

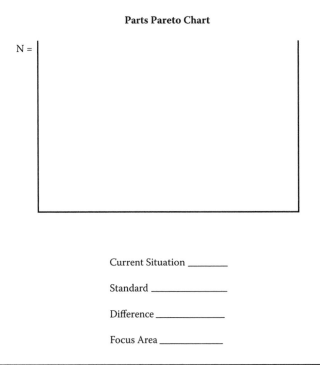

Current Situation _____

Standard _____

Difference _____

Focus Area _____

Figure 12.16   Blank parts Pareto.

Current Situation _____

Standard _____

Difference _____

Focus Area _____

Figure 12.17 Blank defects Pareto.

- Production = 2,000 units
- Standard = 0.025
- Injection Mold Unit #1:
  - Part 1: 1 Flash, 1 Crazing, 0 Blush
  - Part 2: 1 Flash, 1 Crazing, 1 Blush
  - Part 3: 0 Flash, 0 Crazing, 1 Blush
  - Part 4: 0 Flash, 2 Crazing, 0 Blush

- Injection Mold #2:
  - Part 1: 0 Flash, 0 Crazing, 0 Blush
  - Part 2: 6 Flash, 0 Crazing, 4 Blush
  - Part 3: 1 Flash, 10 Crazing, 2 Blush
  - Part 4: 1 Flash, 35 Crazing, 1 Blush

- Injection Mold #3:
  - Part 1: 1 Flash, 4 Crazing, 0 Blush
  - Part 2: 1 Flash, 3 Crazing, 1 Blush
  - Part 3: 0 Flash, 4 Crazing, 1 Blush
  - Part 4: 0 Flash, 6 Crazing, 1 Blush

See Appendix L for answers to this exercise.

Pie Chart Exercise

Production downtime is on the rise and you need to do something to get the attention of the department supervisors. You decide to create a pie chart showing the percentage of downtime by department and post it in the production area. Analysis of last month's overtime shows 2 hours downtime in welding, 7 hours downtime in machining, 3 hours downtime in assembly, 10 hours downtime in paint, and 15 hours downtime in stamping. After converting the downtime hours to percentages, draw a pie chart depicting the percentage of downtime by department. See Appendix M for answers to this exercise.

Pictogram Exercise

The shipping department is just starting to track data on shipments. The department wants to be able to tell at a glance how many 5-gallon buckets of paint are shipped each month. In January, 3,000 buckets were shipped; in February, 2,000 buckets were shipped; in March, 4,000 buckets were shipped; in April, 4,000 buckets were shipped; in May, 6,000 buckets were shipped; and in June, 3,000 buckets were shipped. Draw a pictogram of the paint shipments by month. See Appendix N for answers to this exercise.

Titan Case Part A: A3 with Visuals Practice

August 6, 2010. You are Christopher, the Sales Manager for Titan Tank and Vessel located in Denver, Colorado. Titan has offered premium-quality tanks and vessels since it opened in May of 1962. Every tank and vessel is a one-of-a-kind product as Titan designs each tank based on the exact needs of every customer. Titan prides itself on superior service, unparalleled quality, and delivery speed second to none.

Titan has two support staff and four traveling sales staff covering the states of New Mexico, Colorado, Wyoming, Montana, Idaho, Utah, Arizona, California, Nevada, Oregon, and Washington. The two in-house sales and marketing staff schedule appointments for the four traveling sales staff and create the customer quote. Titan tanks and vessels are used in numerous industries:

■ Agricultural
■ Medical
■ Paper Industry
■ Breweries
■ Distilleries

You track customer complaints by salesperson and type of complaint. Last month, your Customer Service department received a higher number of complaint calls from customers. Normally, only 4% of calls are from customers complaining about products or service. There were 300 calls last month and 23 were complaint calls. As the Sales Manager, you are extremely concerned by the large number of complaint calls. You are not accustomed to this many complaints in one month. Titan takes pride in its ability to produce and deliver the highest-quality tanks and vessels to its customers.

Titan holds quarterly department review meetings in which department managers present the results of their department metrics. The next quarterly review meeting is scheduled for the first week of October. At that time a report is made to Titan's Executive staff. You hope to be able to report that customer complaints have dropped back down to 4% or less by that meeting.

The following table shows the total number of incoming calls to the Customer Service department and the number of calls that were complaint calls for April, May, and June.

Month	Complaint Calls	Total Calls	Complaint Percent
April	12	300	4.0
May	13	317	4.1
June	20	312	6.4

In July there were 300 calls to the Customer Service department. The breakdown of the 23 complaint calls was as follows:

Problem Responsiveness (PR): Complaints based on the customer's perception of how well Titan sales reps respond to issues, problems, and concerns

Delivery Delays (DD): Complaints due to Titan missing the promise date for order delivery

Quality Rating (QR): Complaints about the quality and durability of Titan products

Invoice Accuracy (IA): Complaints about additional costs not identified when quoted

- Evelyn (12 years of experience): Sales Rep. #005:
  - PR = 3
  - DD = 0
  - QR = 2
  - IA = 0

- Bob (10 years of experience): Sales Rep. #006:
  - PR = 3
  - DD = 1
  - QR = 0
  - IA = 0

- Jim (new hire): Sales Rep. #007:
  - PR = 0
  - DD = 6
  - QR = 0
  - IA = 2

- Billy (new hire): Sales Rep. #008:
  - PR = 0
  - DD = 4
  - QR = 0
  - IA = 2

Instructions: Using the information in Titan Case Part A, complete the Theme, Problem Situation, and Target blocks of Appendix Q. To help tell the story, incorporate visuals in your A3.

Titan Case Part B: A3 with Visuals Practice

You came up with three potential causes as to why customers do not get orders as promised: shipping errors, manufacturing delays, and poor scheduling. Additional facts include the following:

- New Titan employees do not receive any product-specific training other than what is learned on the job. Jim and Billy started with the company in April after the two most senior sales reps retired. New sales reps are paired with an experienced sales rep for 1 month to learn the sales process.
- Titan promises customers that delivery will be made 30 days from the date of sale. New sales rep orders did not reach the customer until after the promise date. The new sales reps made their first sales the first week of May but customers did not receive the order until the last week of June.
- All items are shipped to the customer and arrive within 2 days of the unit completion.
- Titan sales reps use a standard checklist to gather all the information about customer requirements. The new sales reps' checklists lack information needed by the manufacturing department to complete the unit in time to meet customer promise dates.
- Production Control records indicate that all units are scheduled to begin production within 48 hours of a customer signing the quote.
- Senior sales reps train new sales reps on how to make appointments and take them on sales calls so that they can get a feel for the types of customers and product uses.
- Manufacturing had to make several inquires about the new sales reps' orders, thus delaying production. New sales reps had to relay Manufacturing's questions to the customer in order to complete the orders.
- When the new sales reps were questioned about the lack of information on the checklist, they stated that they did not feel knowledgeable enough about the Titan products to ask customers the right questions. They felt that the checklist assumes a certain level of product knowledge.
- When questioned about the standardized checklist, the sales reps confirmed that it takes a certain level of product knowledge to accurately and completely fill in the checklist. Their perception was that it is part of the learning curve for a Titan sales rep.

> **Instructions: Using the information from Titan Case Parts A and B, complete the Cause Analysis, determine what Countermeasure to implement, create an Implementation plan, and complete the Follow-up section of Appendix Q. After completing Appendix Q, you may compare your A3 to the example in Appendix R, Titan A3 Answer.**

APPENDICES

A. Production Standards Exercise Answers
B. Office Standards Exercise Answers
C. Production Problem Statement Exercise Answers
D. Office Problem Statement Exercise Answers
E. Target Statement Exercise Answers
F. Theme Statement Exercise Answers
G. Loud-&-Clear Answer
H. Production 5-Why Exercise Answers
I. Office 5-Why Exercise Answers
J. Line Graph Exercise Answer
K. Data Table Exercise Answer
L. Pareto Chart Exercise Answer
M. Pie Chart Exercise Answer
N. Pictogram Exercise Answer
O. Dave's Fabrication Blank A3
P. Dave's Fabrication A3 Answer
Q. Titan Blank A3
R. Titan A3 Answer
S. 5S Problem Report A3
T. Food Service Proposal A3
U. Maintenance Training Proposal A3
V. Blank Problem-Solving A3 Format

Appendix A: Production Standards Exercise Answers

Example 3.1 It should only take a couple of minutes to tap the holes for the door hinges.

The word "couple" to most people means two; however, in problem solving, eliminating misunderstanding is critical. Writing 2 minutes versus a couple minutes eliminates misunderstanding on the part of the reader (Standard level).

Example 3.2 Team Members working the fill station must fill 100 bags of feed every hour .

Good example: Specific in the number of bags to be filled in 1 hour. If you fill 105 bags in an hour, you are doing better than expected. If you fill 99 bags in an hour, you are not meeting the expectation (Standard level).

Example 3.3 All Team Members on the 5S Kaizen team should participate in meetings.

"All Team Members" is specific but what is meant by "participate"? You would have to define for the reader what participation would look like during a 5S Kaizen (Standard way).

Example 3.4 Team Members must place Kanbans in the Kanban mailbox as soon as the first part is pulled from each box.

Good example: It describes how the worker should perform his or her job. The worker knows that the Kanban must be pulled and placed in the Kanban mailbox when the first part is pulled (Standard way).

Example 3.5 Cartridges should slide into brass collars without having to be tapped into position.

Good example: If the worker has to tap the cartridge in any way to get it into the collar, there is a problem with fit (Standard way).

Example 3.6 All welds on the forklift frame must be continuous with no gaps and be between 11 and 11.5 millimeters wide.

Good example: If the welds on the forklift frame are 11 millimeters wide but there is one break in the weld causing a gap, it is not acceptable. If the weld is

10.5 millimeters wide or 12 millimeters wide and continuous, it does not meet the expectation (Standard way—no gaps, Standard level—size 11–11.5mm).

Example 3.7 The height of the label should fall between 30 millimeters and the upper limit of the standard.

The lower limit is 30 millimeters but it does not specify how far beyond 30 millimeters the label can vary. A better example would be 30 to 32.5 millimeters. This lets the reader know the size range for acceptable labels (Standard level).

Example 3.8 It should only take a little while to change from one product packaging to the next.

How long is "a little while"? Be specific regarding the amount of time it should take to change product packaging: 10 minutes, 30 minutes, or 1 hour (Standard level).

Appendix B: Office Standards Exercise Answers

Example 3.9 The Mailroom Team Member should fill all copiers with paper before leaving at the end of each day.

This may seem like a good example; however, the intent is that the Team Member fills the copiers between 4:30 p.m. and 5:00 p.m. The way the standard is written, the reader could assume that it is OK to fill the copiers before lunch. Lunch is before the end of the day (Standard way).

Example 3.10 On the second Monday of every month, all expenses from the previous month should be submitted to the Finance department to be paid on the last Friday of the current month.

Good example: The reader knows that if an expense from the previous month is submitted on the second Tuesday of the month, then it will not be paid by the last Friday of the current month (Standard way).

Example 3.11 Office staff should conduct regular Quality Circle meetings.

What does "regular" mean? Does it mean once per week, once per month, or once per quarter? (Standard level). The interval of meetings must be specified.

Example 3.12 Payroll staff must have all hours for the previous week entered into the Payware program no later than noon every Monday, or by noon on the first day back after a Monday holiday.

Good example: It is clear to the reader that if all payroll hours are entered into Payware by 12:30 p.m. on Monday, it is not acceptable (Standard way).

Example 3.13 The 6-month calendar must be approved and posted on all bulletin boards by the last Wednesday in June and the first Wednesday in December.

Good example: Although specific dates are not used, it is clear to the reader when the calendars are to be posted. If the calendar for the second half of the year is posted on the last Thursday of the month, it is not acceptable (Standard way).

Example 3.14 The receptionist should never leave the front desk unattended for more than a few minutes at a time.

What is "unattended"? How long is "a few minutes"? When pressed for something more specific, the Vice President of Administration said, "A visitor should not have to wait more than a minute for assistance, and the phone should never ring more than three times" (Standard level).

Example 3.15 Software can only be added to company PCs by the Information Systems department.

Good example: This is very specific. If a worker other than an Information Systems worker downloads software to his or her work PC, it is unacceptable (Standard way).

Example 3.16 Personal radios should not be turned up too loud.

What is "too loud"? A more specific standard would be to state, "If the radio can be heard by others outside of your work space, the radio is too loud." (Standard level).

Appendix C: Production Problem Statement Exercise Answers

Example 3.17

- Frame weld must produce a completed head guard every 25 minutes.
- 92% of the time, frame weld produces a completed head guard every 25 minutes.
- 8% of the time, frame weld takes 27 to 31 minutes to complete a head guard.

The information contained in this example describes the standard and the current situation. The discrepancy must be calculated by comparing the current situation to the standard. The discrepancy: 8% of the time, it takes 2 to 6 minutes longer to complete a head guard.

Example 3.18

- It should only take a couple of minutes to tap all three door hinge holes.
- It is taking 10 to 16 minutes to tap all three door hinge holes.
- This is five to eight times longer than expected.

The standard should be clarified (2 minutes versus a couple). The current situation clearly communicates how long it does take to tap all three hinge holes. The discrepancy uses a different form of measurement than the standard, so it must be clarified. If the standard is 2 minutes, then the discrepancy: 8 to 14 minutes longer to tap all three hinge holes.

Example 3.19

- No more than 2% of faucet handles should need rebuffing.
- 5% of gold faucet handles and 3% of silver faucet handles need to be rebuffed.
- Gold handles exceed the standard by 3% and silver faucet handles exceed the standard by 1%.

Good example: This example does a good job of quantifying the problem and breaking it down into gold and silver faucet handles.

Example 3.20

- No more than three injection-molded parts should have color mix after a color changeover.
- Injection mold machines 1, 2, and 4 experience two or three parts with color mix after a color changeover, whereas injection mold machines 5 and 6 experience five to seven parts with color mix after a color changeover.
- Injection mold machine 3 is offline.

This example does a good job of stating the standard and communicating the current situation for all six molds. It does not specify the discrepancy. The discrepancy: molds 5 and 6 have two to four parts too many with color mix after a color changeover.

Appendix D: Office Problem Statement Exercise Answers

Example 3.21

- All 10 supervisors must have all their Team Member performance appraisals submitted to HR no later than April 15.
- As of April 15, 78 of 96 performance appraisals were turned in to HR.
- Performance appraisals are late.

This example has a clear standard describing what must be done by a specific date. The current situation is clear about how many performance appraisals were turned in on time but it could be more specific if it stated the current situation for each supervisor. The discrepancy clearly states that appraisals are late, but not how many are late (18).

Example 3.22

- All ink cartridge boxes are to be kept for shipment to the recycler.
- There are seven cartridges that need to be shipped to the recycler and only four boxes.
- Three ink cartridge boxes are missing.

Good example: How many ink cartridge boxes should be kept? (ALL)
How many cartridges need to go back to the recycler? (7)
How many ink cartridge boxes are there? (4)
The difference between how many are needed (7) and how many there are (4): there are three (3) missing ink cartridge boxes.

Example 3.23

- Only a limited number of staff members can be off at one time.
- Three purchasing staff members are off this week.
- Other departments are complaining about poor service.

The standard needs to specify what is meant by "limited." The current situation is clear about how many are off, but because we are not sure of how many can be off, it is not possible to determine if there is a problem. However, we do know that other departments are complaining about poor service from the department. What needs to be clarified is what is meant by "poor service": Are they waiting too long? Are they not getting the right answers to their questions?

Example 3.24

■ The Safety department set a standard requiring all Team Members to complete one online safety training module per month.
■ 100% of the office and 75% the shop have completed one module per month for the past 3 months.
■ 25% of Shop Team Members do not see the need for the monthly online safety training.

This could be a good example if the discrepancy avoided the assumption that Shop Team Members do not see the need for the monthly online safety training.

Appendix E: Target Statement Exercise Answers

Example 4.1

A. __G__ Increase Safety Committee attendance by 10% before March 25, 2010.
B. __B__ Increase Safety Committee attendance of Shop Team Members by 10% before March 25, 2010.

Both examples provide all four parts of a target statement: Do What (Increase), To What (Safety Committee attendance), How Much (10%), and By When (March 25, 2010). However, target statement "B" provides a better understanding of the characteristics of the problem (Shop Team Members).

Example 4.2

A. __N__ Reduce the time it takes to tap all left side door hinge holes from 15 minutes to 10 minutes by the end of the week.
B. __N__ Reduce the time it takes to tap all left side door hinge holes by August 30, 2010.

Target statement "A" needs a more specific date and "B" needs the amount specified.

Example 4.3

A. __G__ Eliminate downtime on sewing line #4 by no later than November 7, 2010.
B. __B__ Eliminate downtime on sewing line #4 due to broken needles by November 7, 2010.

Example B includes the reason for the downtime (i.e., broken needles). Both examples use the word "eliminate" making it unnecessary to indicate "How Much."

Example 4.4

A. __G__ Improve turnaround time on book binding from 2 weeks to 3 days by April 30, 2010.

B. __B__ Reduce book binding turnaround time from 2 weeks to 3 days by April 30, 2010.

The word "Reduce" is a better representation of what the author of the A3 is trying to accomplish. Using words like decrease or increase provide directionality.

Example 4.5

A. __N__ Increase Team Member morale rating on the 2012 opinion survey.

B. __G__ Improve Team Member morale rating from 80% to 95% on the 2012 opinion survey.

Example "A" does not specify how much of an increase is needed on the survey. Example "B" would be better if the word "increase" were used in place of "improve."

Example 4.6

A. __N__ Move to #95 on the list of best places to work in the state by 2010.

B. __G__ Improve Lotta-Lift's ranking on the list of best places to work in the state from #111 to #95 by May 10, 2010.

The "How Much" and "By When" need to be more specific in example "A".

Appendix F: Theme Statement Exercise Answers

Example 5.1

- Target: Increase Safety Committee attendance of Shop Team Members from 30% to 50% before March 25, 2010.
- *Theme: Increase Safety Committee attendance of Shop Team Members.*

Example 5.2

- Target: Reduce the time it takes to tap left side door hinge holes from 15 minutes to 10 minutes by August 30, 2010.
- *Theme: Reduce the time it takes to tap left side door hinge holes.*

Example 5.3

- Target: Eliminate downtime on sewing line #4 due to broken needles by November 7, 2010.
- *Theme: Eliminate downtime on sewing line #4 due to broken needles.*

Example 5.4

- Target: Reduce book binding turnaround time from 2 weeks to 3 days by April 30, 2010.
- *Theme: Reduce book binding turnaround time.*

Example 5.5

- Target: Increase Team Member morale rating on 2010 opinion survey from 80% to 95% by March 24, 2012.
- *Theme: Increase Team Member morale rating on 2010 opinion survey.*

Example 5.6

- Target: Improve Lotta-Lift's ranking on the list of best places to work in the state from #111 to #95 by May 10, 2010.
- *Theme: Improve Lotta-Lift's ranking on the list of best places to work in the state.*

Appendix G:
Loud-&-Clear Answer

Theme:
Reduce 8- to 10-inch speaker defects due to suspension separation from the diaphragm on blue shift.

Problem Situation:
Background: The 8- to 10-inch speaker line produces a unique clear diaphragm made of a plastic-like material that produces a high-quality sound. A new product line using the clear diaphragm will be produced on the line beginning in June. As the Team Leader, I am responsible for ensuring that quality is maintained.
Standard: No more than 1% quality defects on 8- to 10-inch speaker line.
Current Situation: Quality defects on the 8- to 10-inch speaker line are 10% or 100 defects.
Discrepancy: Quality defects on the 8- to 10-inch speaker line exceed the standard by 9%, or 90 defects. The defects are due to suspension separation from the diaphragm.
Extent: This has been a problem on blue shift since May 18, 2010.
Rationale: It is important that this problem be addressed in order to meet quality standards. If this problem is not addressed now, it will affect line-off of the new product and will get worse if nothing is done.

Target:

Do What:	Reduce
To What:	8- to 10-inch speaker defects due to suspension separation from the diaphragm on blue shift
How Much:	From 10% to 1% or less
By When:	June 10, 2010

Appendix H: Production 5-Why Exercise Answers

Example 7.1
Problem: 25% of fabricated fireproof storage boxes do not fit in home location.

   __1__   Fireproof storage boxes are ¼ inch too tall, wide, and deep.
   __2__   Associate #3's tape measure is off by ¼ inch.
   __4__   Associate #3's tape measure is not documented by Quality department.
   __3__   Associate #3's tape measure is not calibrated.
   __5__   Associate #3 is using his personal tape measure.

Example 7.2
Problem: 10% of runners on plastic molded parts do not fall onto regrind chute.

   __1__   Robotic arm not aligned with regrind chute.
OR
   __1__   Regrind chute is too narrow.
OR
   __1__   Runner sways back and forth before dropping.

If you had difficulty trying to sequence the statements in Example 7.2, it is because the statements are all potential causes. The statements do not progressively lead to the root cause. Each potential cause would need to be checked to see which one caused the problem.

Example 7.3
Problem: 30% of refrigerator water filter cartridges strip out on second shift.

   __3__   Associate thinks cartridges are too loose.
   __1__   Torque set too high on cartridge air tool.
   __4__   Associate does not understand torque requirements for part.
   __2__   Associate increases air pressure.
   __6__   Torque specification not specified on Job Breakdown sheet.
   __5__   Associate not provided specifications during Job Instruction.

Example 7.4

Problem: 18% of inserts damaged when pressed into part.

   3   Neck on press blocks assembler's view.

   1   Insert not aligned flush with part opening.

   2   Assembler cannot see insert in relation to part opening.

Appendix I: Office 5-Why Exercise Answers

Example 7.5

Problem: Picnic tables have trash stuffed into gaps between boards.

   2   Associates do not want to walk 150 feet to dumpster.

   1   Associates do not put trash in dumpsters.

   3   Associates think it takes too much of their breaks and lunchtime.

Example 7.6

Problem: Interoffice mail not delivered at 9:00 a.m., thus delaying payroll process.

   3   Security was asked to transport associate to Urgent Treatment.

   2   Security did not pick up mail until 8:30 a.m.

   4   Night shift supervisor did not want to take the Team Member to Urgent Treatment.

   1   Mail not sorted until 9:00 a.m. by office staff.

   5   Supervisor did not want to stay late to complete production reports.

Example 7.7

Problem: Network copier jammed.

   1   Paper curls around roller.

OR

   1   Paper is too large.

OR

   1   Paper is feeding at the wrong angle.

OR

   1   Too many pages are feeding at the same time.

The statements in Example 7.7 do not progressively lead to the root cause. Each potential cause needs to be checked to see which one caused the problem.

Example 7.8

Problem: Supply room ran out of blue ink pens.

   5   Supply room replacement not specified when office supply staff went on vacation.

   4   Supply room staff did not notify office manager.

   3   Office manager did not know that blue ink pens were needed.

   2   Purchasing did not receive a purchase order for blue ink pens.

   1   Purchasing did not order blue ink pens.

Appendix J: Line Graph Exercise Answer

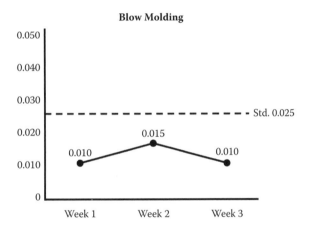

Blow Molding

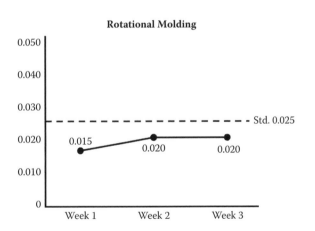

Rotational Molding

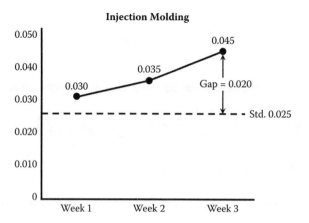

Appendix K: Data Table Exercise Answer

Molding Defect Rate Versus Standard

	Defect Rate	Standard	Gap
Blow Wk 1	0.010	0.025	N/A
Blow Wk 2	0.015	0.025	N/A
Blow Wk 3	0.010	0.025	N/A
Rotational Wk 1	0.015	0.025	N/A
Rotational Wk 2	0.020	0.025	N/A
Rotational Wk 3	0.020	0.025	N/A
Injection Wk 1	0.030	0.025	0.005
Injection Wk 2	0.035	0.025	0.010
Injection Wk 3	0.045	0.025	0.020

Note: Each week the defect rate on the injection machines exceeded the standard. However, the gap is only highlighted on week 3 because it is the difference between the standard and the current situation.

Appendix L: Pareto Chart Exercise Answer

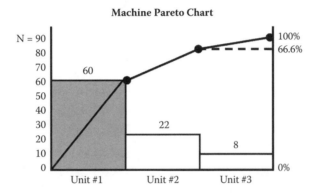

Machine Pareto Chart

Current Situation ___90___

Standard ___50___

Difference ___40___

Focus Area ___60___

Parts Pareto Chart

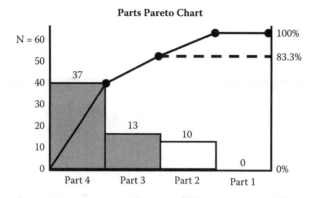

Current Situation _____90_____

Standard _____50_____

Difference _____40_____

Focus Area _____50_____

Defects Pareto Chart

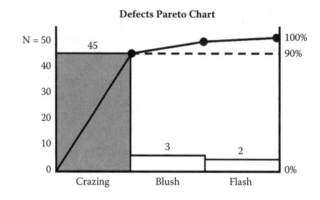

Current Situation _____90_____

Standard _____50_____

Difference _____40_____

Focus Area _____45_____

Appendix M: Pie Chart Exercise Answer

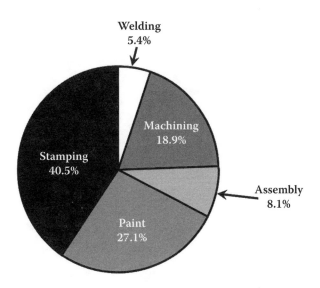

Appendix N: Pictogram Exercise Answer

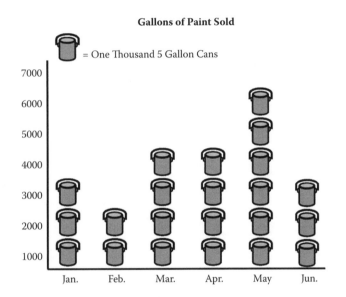

Gallons of Paint Sold

= One Thousand 5 Gallon Cans

Appendix O: Dave's Fabrication Blank A3

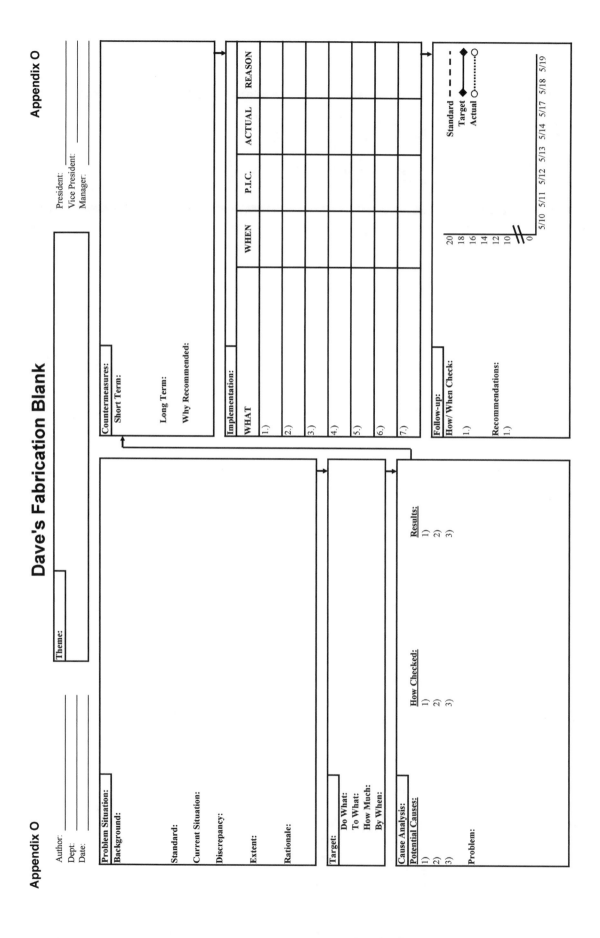

Appendix P: Dave's Fabrication A3 Answer

Dave's Fabrication Answer

Appendix P

Author: Kelly M.
Dept: Welding
Date: May 10, 2010

President:
Vice President:
Manager:

Theme: Reduce backrest weld times due to filling gaps from slats being out of square

Problem Situation:

Background: Welding is responsible for the assembly of fabricated parts prior to painting. The Team Leader is responsible for managing the flow through the welding department to ensure customers get their product on time. Lotta-Lift, the company's largest customer, is doubling their weekly order and number of deliveries from 1 to 2 per week.

Standard: Weld time per inch of .25 minutes multiplied by 44 inches or 11 minutes per backrest.

Current Situation: Weld time per inch of .375 minutes multiplied by 44 inches or 16.5 minutes per backrest.

Discrepancy: Weld time is .125 minutes longer per inch multiplied by 44 inches or 5.5 minutes longer per backrest.

Extent: Since 4/26/10 the backrest Welder has been filling gaps between slats and the backrest frame due to slats not being cut squarely at the shear process.

Rationale: It is important that this problem be addressed now in order to meet the customer's new demand rate. This problem must be addressed in the next week if Dave's Fabrication is going to meet the new requirements. The problem will stay the same if nothing is done.

Target:

Do What:	Reduce
To What:	Backrest weld times due to filling gaps from slats being out of square
How Much:	From 16.6 minutes to 11 minutes
By When:	May 17, 2010

Cause Analysis:

Potential Causes:
1) Operator not aligning part correctly
2) Shear blade misaligned
3) Shear squaring arm misaligned

How Checked:
1) Audit operator to Job Breakdown
2) Measure & compare to alignment specs
3) Check arm to see if it is square

Results:
1) Not an issue
2) Meets specs.
3) Not square

Problem: 5.5 minutes longer to weld backrests due to parts not being cut square at shear process
↳ Shear squaring arm misaligned
↳ Material Handler bumped squaring arm with forklift counterweight
↳ Driving forklift in congested process area are not designated for forklifts
↳ Material Handler thinks it makes no sense to use the overhead crane

Root Cause: ↳ Safety concerns never explained to Material Handler

Countermeasures:

Short Term:
1.) Cut all backrest slats using the band saw until the shear is repaired.
2.) Realign shear squaring arm.
3.) Erect a guardrail barrier to prevent vehicle traffic in shear process area.

Long Term: Review Job Instruction with all Material Handlers emphasizing safety key points.

Why Recommended: The short term will prevent vehicles from entering the shear process & make it possible to meet customer demand until the shear is repaired. The long term will make the Material Handler aware of safety concerns regarding vehicles in process areas.

Implementation:

WHAT	WHEN	P.I.C.	ACTUAL	REASON
1.) Instruct Material Handler on safety key points.	5/11/2010	Diane P.	5/11/2010	
2.) Cut all slats on band saw.	5/11/2010	Bill M.	5/11/2010	
3.) Get P/O approved for contractor to repair shear squaring arm.	5/12/2010	Mart M.	5/12/2010	
4.) Write maintenance W/O for guardrail.	5/12/2010	Kelly M.	5/12/2010	
5.) Instruct all M/Hs on safety key points during team meeting.	5/12/2010	Diane P.	5/12/2010	
6.) Erect guardrail on west side of shear area.	5/13/2010	Eve M. & Willie P.	5/17/2010	
7.) Realign shear squaring arm.	5/31/2010	Contractor	5/25/2010	Material availability

Follow-up:

How/ When Check:
1.) Track average daily time required to weld backrests.

Recommendations:
1.) Identify other areas where guardrails should be erected.

Legend: Standard, Target, Actual

20
18
16
14
12
10
0

5/10 5/11 5/12 5/13 5/14 5/17 5/18 5/19

Appendix P

Appendix Q: Titan Blank A3

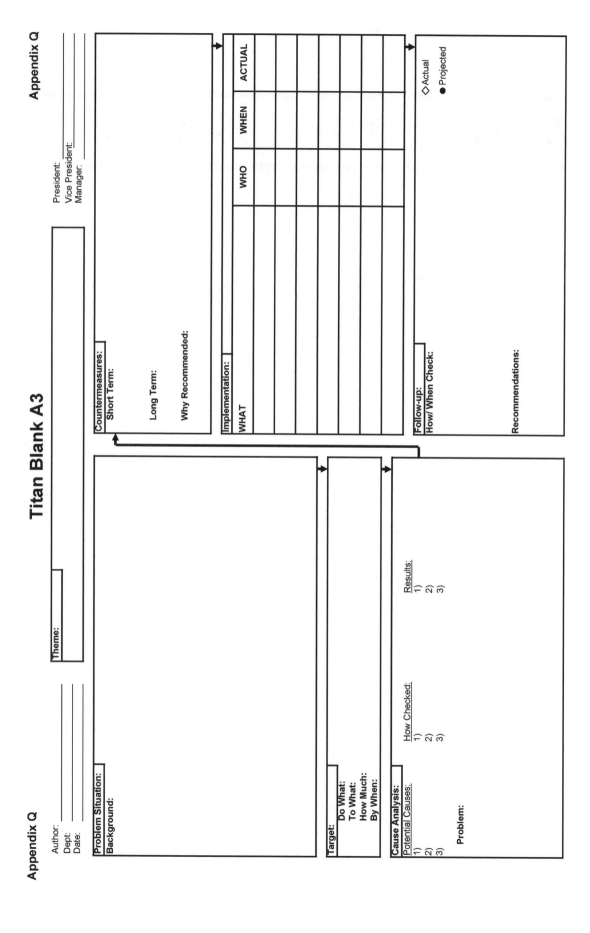

Appendix R: Titan A3 Answer

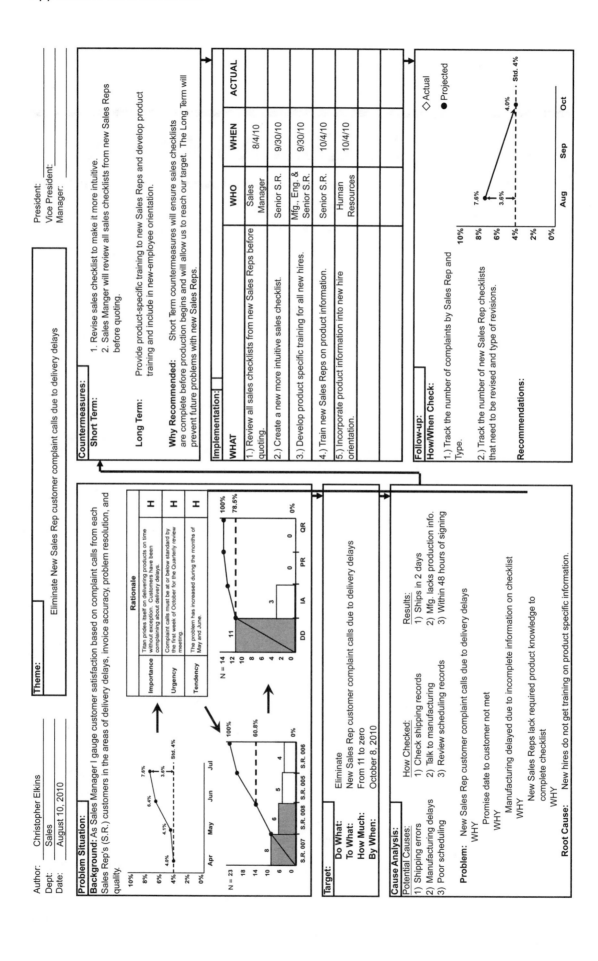

Appendix S: 5S Problem Report A3

Author: Marty Miller
Dept: Coil Line
Date: January 3, 2010

Theme: Eliminate 5S scoring differences between Executive Committee (Team Leaders & Continuous Improvement Champion).

President: _____
Vice President: _____
Manager: _____

Problem Situation: **Background:** I am the Group Leader on the new Coil Line that has only been running production for 5 weeks. In addition to running production we must implement 5S. Due to customer demand there was no time to develop 5S prior to starting production. As a means of motivating areas to improve 5S, any area scoring less than 80 points on quarterly 5S audits must clean the outside common areas.

Expectation: All three audit groups interpret 5S Standards consistently. No more than a 3-point difference between auditors.

December 2009 5S Audit Scores

	Week 1	Week 2	Week 3	Week 4
Team Leader	65	72	79	83
Lean Champion	Monthly Audit			84
Executive Committee	Quarterly Audit			71

Extent: The scoring inconsistency only occurred during the Quarterly Audit with the Executive Committee (E.C.) before 5S could be fully implemented on my line.

Rationale: Team members feel that the E.C. is unreasonable in their expectations for a new line. This problem needs to be addressed immediately in order to keep morale from deteriorating. This could be an issue in other areas of the facility during line starting ups.

Target:
Do What: Eliminate
To What: 5S scoring differences between E.C. & (T/L & C.I.C.)
How Much: From (10 - 15 points) to no more than 3 points
By When: In time for the next quarterly E.C. audit - March 25, 2010

Cause Analysis:
Potential Causes:
1) T/L giving a break in scoring
2) C.I.C. giving a break in scoring
3) E.C. scoring too strictly

How Checked:
1) Analyze T/L method
2) Analyze C.I.C. method
3) Analyze E.C. method

Results:
1) Scored on a curve
2) Scored on a curve
3) Audited to standard

Problem: 10 to 14 percent difference between E.C. & (T/L & C.I.C.) 5S audit scores.
WHY
T/L & C.I.C. scored the new line on a curve.
WHY
T/L & C.I.C. feel it is unfair to expect a new line to meet the 5S standard in such a short period of time while running production.
WHY
Root Cause: Teams can only work on 5S during downtime or during approved overtime.

Countermeasures:
Short Term: Train all auditors in proper scoring to calibrate all three audits. Replace punishment system with a recognition program.

Long Term: Establish 5S on new lines before running production.

Why Recommended: The short term will ensure all 3 audits are conducted fairly and consistently. Removing punishment for low scores will improve morale. Establishing 5S prior to running production will eliminate the need to audit on a curve and provide a better production environment.

Implementation:

WHAT	WHO	WHEN	ACTUAL
1.) Train all 3 audit groups for consistency.	Steering Committee	1/15/10	1/13/10
2.) Test consistency by having all 3 groups audit the same area during the same day.	Steering Committee	1/18/10	1/15/10
3.) After test audit facilitate discussion on audit results with all 3 groups.	Steering Committee	1/18/10	1/15/10
4.) Revise 5S policy to exclude punishment for scores below 80 points & create reward system.	Human Resources	2/19/10	2/19/10
5.) Create rotating multi department teams to clean outside common areas.	Executive Committee	2/26/10	2/19/10
6.) Establish a plan to 5S all new lines prior to running production.	Eng. Dept.	3/12/10	3/12/10

Follow-up:
How/When Check:
1.) Compare quarterly audit results to ensure consistency within 3 points.

2.) Track 5S implementation on new lines.

Recommendations:
Continue to compare quarterly audit results with weekly and monthly 5S audits. If results vary more than 3 points review for other causes.

○ T/L
◇ C.I.C.
● E.C.

3-point range

| 1st Qtr. 2010 | 2nd Qtr. 2010 | 3rd Qtr. 2010 | 4th Qtr. 2010 |

Appendix T: Food Service Proposal A3

Food Service Proposal A3

Author: Rose Miller
Dept.: Human Resources
Date: April 16, 2007

President: _____
Vice President: _____
Manager: _____

Theme:

Improve the quality of food vendor service

Background:

Due to Lotta-Lift's proximity to outside food vendors and the amount of time team members have for breaks and lunch, it is very important that we have a reliable food vendor that offers quality products at a reasonable cost.

Team members rely on the food vending machines before work, on breaks, and during lunch as their only source of food.

Vinny's Vending is the current food vendor and has been the food vendor since 2001.

Lotta-Lift does not receive any commission from vending sales.

Vinny's provides one hot meal per week and has an on-site attendant.

Vinny's provides snack, cold food, and drink machines in all 4 cafeterias.

Vinny's provides tables and chairs in all 4 cafeterias with a total capacity for 300.

Vinny's implemented a price increase two weeks ago without notifying Lotta-Lift.

Problems / Concerns:

Team members complain about the lack of variety offered in the vending machines.

Vending machines are not cleaned and are only 50% stocked.

Prices of items not consistent in every break area (i.e., chips $.40 in the front cafeteria and $.50 in the Warehouse cafeteria).

Drinking fountain and coffee machine in the front cafeteria have insects in the beverages. Several team members have actually drunk from a cup only to discover a bug, still alive, floating in their drink.

Vending machines are outdated and there is no preventive maintenance performed on machines.

Weekly, several machines have mechanical problems that prevent team members from purchasing food items.

Food machines are carousel style and take 1-2 minutes to cycle through. This reduces the amount of time team members have to eat. This does not include the time team members stand in line waiting for others to make their selections.

Analysis:

Human Resources identified and visited 3 different food vendors in addition to Vinny's to assess several criteria.

Rating Scale:
Good = G
Fair = F
Poor = P

Criteria	3 Bros. Vending	Kathy's Catering	Vinny's Vending	F & B Vending
1. Food Quality	F	F	P	G
2. Facility Cleanliness	G	G	F	G
3. Equipment Quality	G	F	P	G
4. Maintenance Program	F	F	P	G
5. Office Coffee Program	G	G	G	G
6. Cafeteria Ability	P	P	P	G
7. Response Time	G	G	G	G
8. Team Member Incentives	P	G	G	G
9. On-Site Staff	G	G	G	G
10. Years in Business	11	5	25	55
11. Satisfaction Rating Excellent = 1 – Poor = 5	2	3	4	1
12. Vending Commission	10%	12%	0%	5%

Recommendations / Benefits:

Improve Lotta-Lift's vending program by switching to F & B vending:
- F & B offers a large variety of good quality, affordable food items.
- F & B has the ability to run a full-service cafeteria.
- F & B will have an on-site attendant working on commission so they will work harder to give better service.
- F & B will run promotions on food vending items.
- F & B will install new equipment reducing the time it takes to make selections.
- Lotta-Lift will receive vending commissions averaging $19,996.00 a month. The money will go to Lotta-Lift's charity fund.
- Better service will improve team member morale.
- No contract will be signed; this will provide F & B the incentive to provide the best service possible.
- F & B has the ability to provide hot lunches once per week in each cafeteria.

Implementation:

	Apr.	May	Jun.	Jul.	Aug.	Sep.	Oct.	Nov.	Dec.
Proposal approved.		15th							
Swap equipment		30th							
Survey employees			30th			30th			20th

Cost:

Vinny's will sell the tables & chairs to Lotta-Lift for $8,000 versus F & B replacing them at a cost of $14,000. Vinny's will accept monthly payments which can be paid using the monthly vending commissions. As a result there will be zero cost.

Appendix U: Maintenance Training Proposal A3

Appendix U

Maintenance Training Proposal A3

Appendix U

Theme: Increase the quality of in-house Maintenance Team Members

Author:	James R. Johnson
Dept.:	Maintenance
Date:	April 3, 2005

President:
Vice President:
Manager:

Current Situation:

Skilled maintenance Team Members (T/M) were hired when the company opened. As turnover in the maintenance department occurred, openings were filled internally by team members.

New maintenance T/Ms are encouraged to participate in the voluntary maintenance training classes in addition to the On-The-Job training and mentoring they receive.

The current program was established in April of 2003.

Currently the company pays $3,500.00 per on-site class that is taught by the local community college. Each class allows up to 12 participants. Typically only 1 or 2 maintenance T/Ms enroll.

Time spent in training is compensated by the Company at the T/M's straight time rate.

There are 4 Modules. At the completion of each module, eligible T/M's receive a $.25 pay increase.

Current maintenance T/Ms can only perform basic maintenance.

Problems / Concerns:

60% of all complicated maintenance tasks have to be contracted out to various maintenance specialists at a cost of $2.5 million a year.

When skilled maintenance are hired from outside the company it creates morale issues.

The one time an external maintenance person was hired it took him 1 year to become familiar with our machines, equipment, and environment.

When an unskilled internal candidate is transferred to maintenance it takes at least 2 years of On-The-Job training to enable them to do basic maintenance and preventive maintenance.

The company pays for 10 to 11 empty seats for each on-site class.

Proposal:

Maintain the Current Policy – The Maintenance Skill Development policy will continue to provide training for current Maintenance T/Ms in the process of completing the maintenance modules. The policy will also remain in place to allow for the hiring and training of external candidates as necessary.

Expanding Maintenance Training – Allow interested T/Ms to fill open slots in Maintenance Modules. Open positions will be filled based on criteria such as seniority, attendance, safety, and quality. These team members will take the courses on a voluntary basis, during unpaid time. The modules will be covered under the company's Tuition Reimbursement program.

Proposal Continued:

Create a Maintenance Pool – After completion of all 4 modules T/Ms will be added to the Maintenance Pool List. This is not a guarantee of transfer.

Transfer Process & Pay Increase - The Maintenance Supervisor will create a proficiency check sheet. Once T/Ms achieve proficiency and have worked in maintenance for 6 months they will receive the $1.00 pay increase.

Maintenance Openings – If openings arise prior to the first group completing all 4 modules, the Maintenance Pool T/Ms will need to complete mentoring activities for the modules they have completed and then will fall under the original policy for the remainder of their training.

Benefits:

- Increases the Quality of Internal Maintenance Candidates.
- Increases Team Member Morale.
- Fills seats that are paid for but would otherwise be empty.
- Creates a formal system for internal T/Ms to transfer into the Maintenance Department.
- Establishes a formal Mentor Program to assist T/Ms in applying their classroom knowledge.
- Reduces annual contracted out maintenance costs.

What	Who	Where	Projected Date	Actual Date	Comments
Get proposal approved	James J.	Corp. Office	4/15/2005	4/13/2005	
Write policy & approve	Bob L.	H. R. Dept.	4/30/2005	4/24/2005	
Announce program	Rose M	Plant Meeting	5/21/2005	5/21/2005	
Graduate first group	Community College	On Site	10/31/2006	10/31/2006	
Eliminate contract maintenance	Maintenance	Maint. Dept.	4/30/2007	12/15/2006	

Cost:

There are no additional costs associated with creating a pool of qualified in-house maintenance T/Ms who can fill open maintenance positions. By filling the on-site classes the company will get more value for the money spent on in-house classes taught by the community college.

Appendix V: Blank Problem-Solving A3 Format

Blank Problem Solving A3 Format

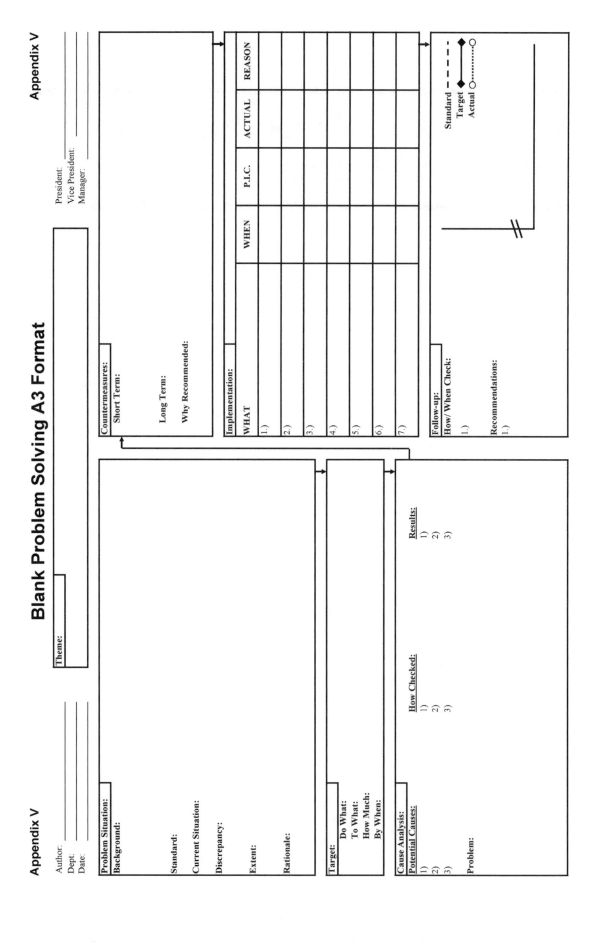

Appendix V

Appendix V

Author: _____
Dept: _____
Date: _____

President: _____
Vice President: _____
Manager: _____

Theme:

Problem Situation:
Background:

Standard:

Current Situation:

Discrepancy:

Extent:

Rationale:

Target:
Do What:
To What:
How Much:
By When:

Cause Analysis:
Potential Causes:
1)
2)
3)

Problem:

How Checked:
1)
2)
3)

Results:
1)
2)
3)

Countermeasures:
Short Term:

Long Term:

Why Recommended:

Implementation:

WHAT	WHEN	P.I.C.	ACTUAL	REASON
1.)				
2.)				
3.)				
4.)				
5.)				
6.)				
7.)				

Follow-up:
How/When Check:
1.)

Recommendations:
1.)

Standard - - - - -
Target ◆——◆
Actual ○·········○

Index